This book is about the atmosphere and h̶... chemistry is probably one of the oldest br... science of air chemistry began as recently ... considerable public interest. This being the ⎯⎯ ...ition of an extremely popular text, much of the book has been rewritten and all information updated, to keep pace with this important and fast moving science.

In the early chapters of his book, Dr Peter Brimblecombe discusses the geochemical, biological and maritime sources of trace gases; these are followed by chapters on the chemistry of atmospheric gases, suspended particles and rainfall. After dealing with the natural atmosphere the book examines the sources of air pollution and its effects: injury to health, damage to plants and animals, degradation of constructional materials, pollution indoors, acidification of rain and changes in global carbon dioxide and methane. These scenarios have been rewritten from the last edition to include recent findings. The final chapter concerned with the chemistry and evolution of the atmospheres of the planets of the solar system has been revised in accordance with current understanding.

Students of chemistry, environmental science, ecology, geography and public he... will ... ...he ... ...erstandi...g of air ch...

D... ... ...r in the School of Environmental Scie... ...st Anglia.

AIR COMPOSITION & CHEMISTRY

CAMBRIDGE ENVIRONMENTAL CHEMISTRY SERIES

Series Editors

Professor P.G.C. Campbell, *Institut National de la Recherche Scientifique, Université du Québec*

J.N. Galloway, *Department of Environmental Sciences, University of Virginia, USA*

R.M. Harrison, *Department of Environmental Health, School of Chemistry, Birmingham University*

AIR COMPOSITION & CHEMISTRY

PETER BRIMBLECOMBE

School of Environmental Sciences, University of East Anglia

Second edition

CAMBRIDGE
UNIVERSITY PRESS

Published by the Press Syndicate of the University of Cambridge
The Pitt Building, Trumpington Street, Cambridge CB2 1RP
40 West 20th Street, New York, NY 10011-4211, USA
10 Stamford Road, Oakleigh, Melbourne 3166, Australia

© Cambridge University Press 1986

First published 1986
Second edition 1996

Printed in Great Britain at the University Press, Cambridge

British Library cataloguing in publication data

Brimblecombe, Peter
Air, composition and chemistry

1. Atmospheric chemistry
I. Title
551.5′11   QC879.6

Library of Congress cataloguing in publication data

Brimblecombe, Peter, 1949–
Air composition and Chemistry / Peter Brimblecombe. – 2nd ed.
    p.   cm. – (Cambridge environmental chemistry series)
Includes bibliographical references and index
ISBN 0 521 45366 6. – ISBN 0 521 45972 9 (pbk.)
1. Atmospheric chemistry. 2. Air – Pollution. I. Title.
II. Series.
QC879.6.B76 1996
551.5′11–dc20   95-7850CIP

ISBN 0 521 45366 6 hardback
ISBN 0 521 45972 9 paperback

CONTENTS

○ ○

PREFACE TO THE FIRST EDITION

○ ○

One could argue that air chemistry is one of the oldest branches of chemistry. The Greek philosopher Anaximenes (*c.* 570–500 BC) believed air to be the fundamental element from which everything else was derived. Certainly the study of air proved a fruitful way for scientists of the early modern period to establish chemical laws. The major gases of the atmosphere (nitrogen, oxygen and carbon dioxide) were all recognised in the eighteenth century. Transient ones such as ammonia, sulphur dioxide and the nitrogen oxides were also detected then, but excited little interest. In the first half of the current century the science of atmospheric chemistry was largely integrated with that of upper atmosphere physics; this aided the development of the science of aeronomy. However, modern atmospheric chemistry began as late as the 1950s. The clearest indications of the directions the science was to take can be found in C. E. Junge's lengthy review in *Advances in Geophysics* of 1958, which expanded into the classic *Air Chemistry and Radioactivity* (1963). This work gave importance to the chemistry of trace gases, aerosols and rainfall, which have all remained central themes of atmospheric chemistry. Atmospheric chemistry is now the focus of considerable public interest and often has to tread the difficult path between being a strictly scientific discipline and an implement for environmental change.

A list of further reading is appended to each chapter; this should enable the reader to pursue the subjects further. I have tried to keep references to specialist scholarly literature to a minimum and to draw attention, where possible, to textbooks and broad reviews. The review literature gives in-depth references, enabling each topic to be explored fully.

I would like to thank all those people who helped me with their time and advice while preparing this book: Simon Clegg, Malcolm Downes, Dr Roy Harrison, Sr Ann Henderson-Sellers, Dr Peter Liss, Frances Nicholas, Dr Martyn Tranter and Dr Andrew Watson. One great debt can never be properly acknowledged. It is to my PhD supervisor, the late Dr John Spedding, who first drew my attention to atmospheric chemistry through a wonderful set of honours lectures at the University of Auckland. To him I owe the opportunity of experiencing the growth of a young science.

Norwich, February 1985   Peter Brimblecombe

PREFACE TO THE SECOND EDITION

○ ○

Almost a decade has passed since I wrote the first edition of this book. With the rapid increase of interest in atmospheric chemistry, enormous advances are hardly surprising. Interestingly, I am often struck by the fact that, in many areas, such as the description of photochemical smog, the framework of the discipline had already been on a very solid footing in 1985. The views held in the 1980s remained largely unchanged. By contrast, the dramatic ozone hole that forms over the south pole each spring has been an indication of the huge changes that we have caused in the atmosphere. Understanding the origin of the ozone hole has involved the development of heterogeneous stratospheric chemistry. These developments were still to come when the first edition was being written.

We have continued to see declines in the concentrations of the traditional pollutants (smoke and sulphur dioxide) in the air of many industrial cities, but interest has spread to a wider range of pollutants. These novel pollutants are often organic materials, some with carcinogenic properties. Fine particles, less than $10\,\mu m$ in diameter, in the urban atmosphere appear to be a particular health hazard. The rising density of automobiles in the larger cities of Europe has led to high concentrations of nitric oxide. These have been so large that a hitherto unimportant reaction, ($2NO + O_2 \rightarrow 2NO_2$), second order in nitric oxide, has become significant in winter pollution episodes. The acid rain issue was of great public interest in the mid 1980s, but is now quite muted. Sulphuric acid deposition has declined somewhat, but that of nitric acid hardly at all. Concern about acid deposition has now shifted to regions of the developing world where emissions are likely to increase.

There has been a growing interest in naturally emitted volatile materials. Perhaps most striking has been the awareness that there are numerous natural sources of halogenated compounds. A decade ago it was thought that most halogenated compounds in the environment were anthropogenic. Now we know that nature is a prolific source. Further afield, Voyager 2 flew by Uranus and Neptune in the second half of the 1980s, completing a great period of planetary exploration by space probes. The Hubble Space Telescope, despite early optical problems, has enabled a continuing study of planetary atmospheres. In particular the recognition of Pluto as a Charon–Pluto system, and fortuitous occultations, have produced unique opportunities to study the farthest reaches of the solar system.

Expanding effort in atmospheric chemistry has ensured that it has become a mature discipline. We can expect many exciting new discoveries in the next decade, but the maturity of the science also means that it is likely to be based on an increasingly stable theoretical footing. We can look forward to generalisable notions, rules and laws becoming widely applicable across more of the subject.

Finally, I would like to thank my colleagues who offered advice during the preparation of the manuscript, most notably Trudy McMullen, Ken Nicholson, Tim Jickells, David Shooter, Chris Porter, Simon Watts, John Green and John Plane. I would also like to acknowledge the inspiration I have received from Boris Greenstreet and his assistant Mr A. Hood, who always know how to remain calm in the face of adversity.

Peter Brimblecombe

NOTE

○ ○

Some of the quantities involved in atmospheric chemistry are very small and others are very large so a range of SI prefixes has to be used. It is important to recall the following:

E, exa	10^{18}	a, atto	10^{-18}
P, peta	10^{15}	f, femto	10^{-15}
T, tera	10^{12}	p, pico	10^{-12}
G, giga	10^{9}	n, nano	10^{-9}
M, mega	10^{6}	μ, micro	10^{-6}
k, kilo	10^{3}	m, milli	10^{-3}

I have adopted SI units throughout the book with only a few exceptions. Time is often expressed as days (d) or years (a). Pressure has been expressed in atmospheres (atm) rather than pascals (Pa) in some cases (1 atm = 101 325 Pa).

1

○ ○

The Atmosphere

1.1   Chemical composition

The composition of the atmosphere has fascinated natural philosophers from the earliest times. The ancient Greeks regarded air as one of the four elements. It was not until the seventeenth century that air was recognised as a mixture of gases. The English chemist Robert Boyle thought this to be so when he wrote in the seventeenth century that air was 'a confused aggregate of effluviums'. Although later, even after nitrogen and oxygen had been observed as the principal components of air, the question of whether it was a mixture remained. The famous English scientist Sir Humphrey Davy (1778–1829) thought it to be a compound. The reason that this belief survived so long was that scientists thought that, if it were not a compound, then the heavier gas oxygen should sink below nitrogen and thus oxygen should be found at higher concentrations at the very bottom of the atmosphere. The strength of the mixing process, which would destroy this gravitational separation, was not appreciated. However, air is a mixture, and traditional reasons for believing this are:

   (i) the ratio of oxygen to nitrogen does vary, just slightly, from place to place;

  (ii) if it were a compound, the formula would be $N_{15}O_4$, which seems unlikely;

 (iii) the physical properties of air are identical to those of the appropriate mixture of nitrogen and oxygen;

 (iv) it is possible to separate the nitrogen and oxygen; and

  (v) there is no volume change or heat release on mixing oxygen and nitrogen.

Accurate analyses of air were made by Henry Cavendish in the 1780s. The remarkable thing about his painstaking analyses was that, no matter how hard he tried to combine all the oxygen and nitrogen in the air chemically, a small inert fraction was always left over. Unknowingly he had discovered the presence of the inert gas argon, which was not isolated until the work of Rayleigh and Ramsay at the end of the 19th century. Table 1.1 lists the concentrations of the major components of the atmosphere. These are the relatively persistent gases that act as the background matrix within which the intricate chemistry of the atmospheric trace components takes place.

It is important here to mention the units of concentration used in atmospheric chemistry. In Table 1.1 we see two expressions for concentration: percentages and ppm (parts per million). Both these units are 'by volume', so 1 ppm is equivalent to a cubic centimetre of gas in a cubic metre of air. From Dalton's Law of partial pressure it is known that the fractional pressure of a gas depends on the fractional amount of the gas present. Thus at ground level, where the pressure is one atmosphere (i.e. 1 atm = 101 325 Pa), a gas present at 1 ppm concentration will have a pressure of 10^{-6} atmospheres. Percentages will be used very rarely beyond this chapter, because most of the gases in which we are interested are present in only trace amounts.

Using concentration units that are 'by volume' is convenient, because there is a direct relationship between the volume or partial pressure of a trace gas and the number of molecules present. Thus for a gas, such as methane, found in air at about 1 ppm, one in every million molecules will, on average, be a molecule of methane. When gases are present in extremely low concentrations it may be necessary to refer to the number of molecules of the gas per unit volume. In ambient air at atmospheric pressure there are 2.69×10^{19} molecules in every cubic centimetre (this is sometimes called the Loschmidt number). A gas present at 1 ppm will

Table 1.1. *Composition of dry unpolluted air by volume*

Nitrogen	78.084%
Oxygen	20.946%
Argon	0.934%
Carbon dioxide	360 ppm (variable)
Neon	18.18 ppm
Helium	5.24 ppm
Methane	1.6 ppm
Krypton	1.14 ppm
Hydrogen	0.5 ppm
Nitrous Oxide	0.3 ppm
Xenon	0.087 ppm

have a concentration of 2.7×10^{13} molecules cm^{-3} in these units.

Condensed phases, i.e. solid and liquid materials, are expressed in terms of mass per unit volume of air, usually in the units $\mu g\,m^{-3}$ $(10^{-6}\,g\,m^{-3})$ i.e. micrograms per cubic metre of air. This is a logical choice as a unit for particles in the atmosphere, but it is sometimes used for gases, for which it becomes a little clumsy. This is because different molecules have different masses, so although two gases have the same concentration in mass terms they may be at quite different concentrations in molecular or pressure terms. There is no general conversion factor between a unit such as ppm and $\mu g\,m^{-3}$ because it is necessary to allow for the molecular weight of the gas under consideration. Some conversion factors that incorporate molecular weight are listed in Table 1.2. However even this presents difficulties because some writers will use the unit $\mu g\,m^{-3}$ of a gas such as nitrogen dioxide to refer to the mass of nitrogen and not the mass of NO_2 that is about three times greater. The issue is confusing although it can be resolved by writing the concentrations as $\mu g(N)\,m^{-3}$ or $\mu g(NO_2)\,m^{-3}$. In this way it is obvious whether the weight refers to elemental nitrogen or the compound. Where possible, this book will try to avoid confusion by using volume and pressure units for gas phase concentrations.

Table 1.1 shows that the atmosphere is predominantly nitrogen and oxygen. Table 1.3 compares its elemental composition with that of other

Table 1.2. *Conversion factors for converting parts per million to $\mu g\,m^{-3}$ at $0\,°C$. The weight is that of the total molecule, except where marked otherwise. These conversion factors can be corrected for different temperatures by multiplying K by 273/T, where T is the absolute temperature (from Junge (1963)).*

Gas	Conversion factor, K $c\,(\mu g\,m^{-3}) = Kc\,(ppm)$
Hydrogen (H_2)	89
Helium (He)	178
Methane (CH_4)	712
Carbon monoxide (CO)	1259
Ozone (O_3)	2140
Nitrous oxide (N_2O)	1960
Nitrogen dioxide (NO_2)	2050
Nitrogen dioxide (NO_2) as N	625
Ammonia (NH_3)	760
Sulphur dioxide (SO_2)	2860
Hydrogen sulphide (H_2S)	1520

major reservoirs on the Earth. The atmosphere is unique in containing significant quantities of the noble gases that are inert to most chemical reagents. Nitrogen is not particularly reactive and may be regarded as a semi-inert element. A remarkable feature of atmospheric composition is the similarity between the atmosphere and the biosphere. It is often said that seawater resembles the composition of the living organisms that originated within it, and it is true that the seas and life share the hydrogen and oxygen atoms of water as their principal components. However perhaps even more significant is the fact that the four principal components of living material are to be found among the first five major components of the atmosphere. Perhaps this suggests strong links between the atmosphere, the biosphere and the origin of life.

1.2   Residence time

The chemistry of the atmosphere is treated in the above section in a static way. It is necessary to approach atmospheric chemistry from a more dynamic point of view, because trace components are constantly produced within the atmosphere or released by sources at the surface of the Earth. Despite this continual addition of material, the overall composition of the atmosphere does not change very much on long timescales. This is because there are sinks that remove the gases from the atmosphere and act to balance input and output. Such a balance maintains the atmosphere in a steady state and its composition exhibits an apparent constancy.

The steady state situation can be imagined in terms of an overflowing tub of water. If the tub is full and flow into it is maintained, then the inflow will equal the outflow. An important parameter describing such a steady state situation is the residence time of material in the system. In the example of a tub this will be equal to the volume of the tub divided by the rate of inflow (or outflow). Alternatively, this can be written as

$$F_i = F_o = A/\tau \tag{1.1}$$

Table 1.3. *Rank order of the elements in terms of the number of atoms in the major reservoirs*

Biosphere	Atmosphere	Hydrosphere	Crust	Mantle	Core
H	N	H	O	O	Fe
O	O	O	Si	Si	Ni
C	H	Cl	Al	Mg	C
N	Ar	Na	Fe	Fe	S
Ca	C	Mg	Mg	Al	Si

where F_i and F_o are the flux of substance into and out of the system, A the amount of substance in the reservoir and τ is the residence time of the substance in the system. Nitrogen has a very long residence time in the atmosphere, probably as much as a million years. Oxygen is more reactive and has a residence time of about 5000 a (a = year), but the noble gas argon has a residence time of 10^7 a. Carbon dioxide and water have much shorter residence times of 4 a and ten days respectively.

The residence time characterises a system only when it is well mixed. If well mixed the probability of removal of material at the sink region is constant. This means that the average age, τ_a, of the molecules of the substance will be identical to the residence time, τ. However one can imagine situations in which this would not occur. Take, for instance, a reservoir with a slow rate of mixing, and source and sink regions isolated from each other. This is the case when particulate materials, which can only be removed in the lower atmosphere, are injected high into the stratosphere. Here the average age of particles is less than the residence time. The opposite case, for which the residence time is shorter than the average age is found when the source and sink are close together, whereby material will be removed very rapidly. This may occur with sulphur dioxide emitted near ground level. Some of this material may be rapidly absorbed at the surface, while that higher up in the atmosphere may have a much longer lifetime. One way of overcoming the problem of treating poorly mixed reservoirs is to divide the system up into several smaller reservoirs. Thus, in the previous example one could imagine a reservoir for the sulphur dioxide close to the ground and another for the longer-lived sulphur dioxide higher up in the atmosphere.

1.3 Size and pressure variation

The air offers so little resistance to our everyday activities that it is often hard to appreciate that it has a density of around $1.2 \, \text{kg m}^{-3}$. Naturally the entire atmosphere has an enormous mass, but even so it is one of the least massive of the reservoirs considered within global geochemistry. Another important characteristic is the extreme mobility of the atmosphere. This is also shown in Table 1.4, which compares the mass and mixing of the atmosphere with those of the other major reservoirs of the Earth. The atmosphere appears to mix very rapidly. Here we mean mixing within a given reservoir such as the atmosphere rather than the rate at which it exchanges with other reservoirs. Thus, a parcel of a gas such as water, with a residence time of only ten days, will not mix with a large fraction of the atmospheric mass. This is because the general mixing time of the atmosphere is about 80 days. On the other hand, carbon dioxide has a much longer residence time and should be well mixed.

In major reservoirs gravity can lead to large pressure variations with depth. In the atmosphere the pressure change alters the density of gases. The variation of pressure with altitude can be derived from the gas laws. In a very thin layer of gas, the rate of change of pressure across the layer will be

$$\mathrm{d}p/\mathrm{d}z = - g\rho \tag{1.2}$$

where p is pressure, g the acceleration due to gravity, ρ density and z altitude. The density of a gas is related to pressure by the equation

$$p = \rho R T / M_m \tag{1.3}$$

where R is the universal gas constant ($8.314\,\mathrm{kmol}^{-1}\,\mathrm{K}^{-1}$), T the absolute temperature and M_m the mean molecular weight (28.966). Substituting this equation into (1.2) gives

$$\mathrm{d}p/p = \frac{- g M_m \,\mathrm{d}z}{RT} \tag{1.4}$$

If the mean molecular weight and temperature are known with altitude then (1.4) can be integrated, setting p_0, the pressure at the surface, as a constant of integration:

$$p_z = p_0 \exp\left[- g M_m z / (RT) \right] \tag{1.5}$$

The exponent can be simplified by substituting H for $RT/(gM_m)$ to give the form normally known as the barometric law.

$$p_z = p_0 \exp\left(- z/H \right) \tag{1.6}$$

Table 1.4. *Size and general mixing of various reservoirs. Plants, animals and organic matter are included in the biosphere, but not coal or sedimentary carbon.*

	mass (kg)	mixing time (a)
Biosphere	4.2×10^{15}	60
Atmosphere	5.2×10^{18}	0.2
Hydrosphere	1.4×10^{21}	1600
Crust	2.4×10^{22}	$>3 \times 10^7$
Mantle	4.0×10^{24}	$>10^8$
Core	1.9×10^{24}	

H is termed the exponential scale height. This useful parameter represents the change in height required to achieve a pressure drop by a factor of $1/e$. It is also the thickness of the entire atmosphere if it were compressed so as to have the density at ground level throughout its entire depth. The form of the equation is similar for the variation with height of the number of molecules per unit volume, n. The parameters n_z and n_o replace the pressure terms. The scale height and some other properties of the atmosphere at ground level are listed in Table 1.5.

The profile of the Earth's atmosphere illustrated in Fig. 1.1 shows just how rapidly pressure falls off with altitude. The decrease is smooth and monotonic, such that structure is apparently absent. Yet each of the reservoirs listed in Table 1.4 has fine levels of structure. In the atmosphere, this fine structure is not always as obvious as it is with the stratification of rocks within the Earth's crust, it is nevertheless important. In the fluid reservoirs, temperature changes are often the most important structural indicators. The interaction of radiation with the atmosphere leads to the thermal structure illustrated in Fig. 1.1.

1.4   Radiation and thermal structure

In Fig. 1.2 is shown an idealised black body spectrum of the sun that emits radiation at a temperature of about 6000 K. The sun does not behave as a perfect black body. There are spectral emissions from electronic transitions, so the actual spectrum is not as continuous as that of a black body. Besides this, absorption of light by the outer photosphere of the sun causes the intensity in the far ultraviolet to fall more rapidly than would be expected for a black body at 6000 K. Temperatures as high as 10^6 K are characteristic of the chromosphere and the corona of the sun (the white halo visible in eclipses). Emissions from these regions are responsible for the extreme ultraviolet and X-ray parts of the solar spectrum.

The temperature of a planet irradiated by solar radiation can be

Table 1.5. *Standard properties of the atmosphere at sea level*

Density	$1.2250014 \, \mathrm{kg \, m^{-3}}$
Gravitational acceleration (g)	$9.80665 \, \mathrm{m \, s^{-2}}$
Kinematic viscosity	$1.4607 \times 10^{-5} \, \mathrm{m^2 \, s^{-1}}$
Mean free path	$6.632 \times 10^{-8} \, \mathrm{m}$
Molecular weight (M_m)	28.966
Number density (n)	$2.5476 \times 10^{19} \, \mathrm{cm^{-3}}$
Pressure (p)	101325 Pa
Scale height (H)	8434 m

estimated by balancing the amount of incoming radiation absorbed, R_a, against the outgoing radiation, R_o. The amount of radiation absorbed by the Earth will be the product of the solar irradiance, I, the area of the Earth and the fraction of the light that is absorbed. The albedo, A, represents that fraction of light that is reflected, so the fraction absorbed will be $1 - A$. The area of the Earth that we require is not the total area, but rather the area as seen by the incoming radiation, which might be

Fig. 1.1. The vertical structure of the Earth's atmosphere. If the scale height, H, were independent of height then this semi-logarithmic graph would show pressure as a straight line. The curvature arises through variation in temperature, mean molecular mass and gravitational attraction with altitude.

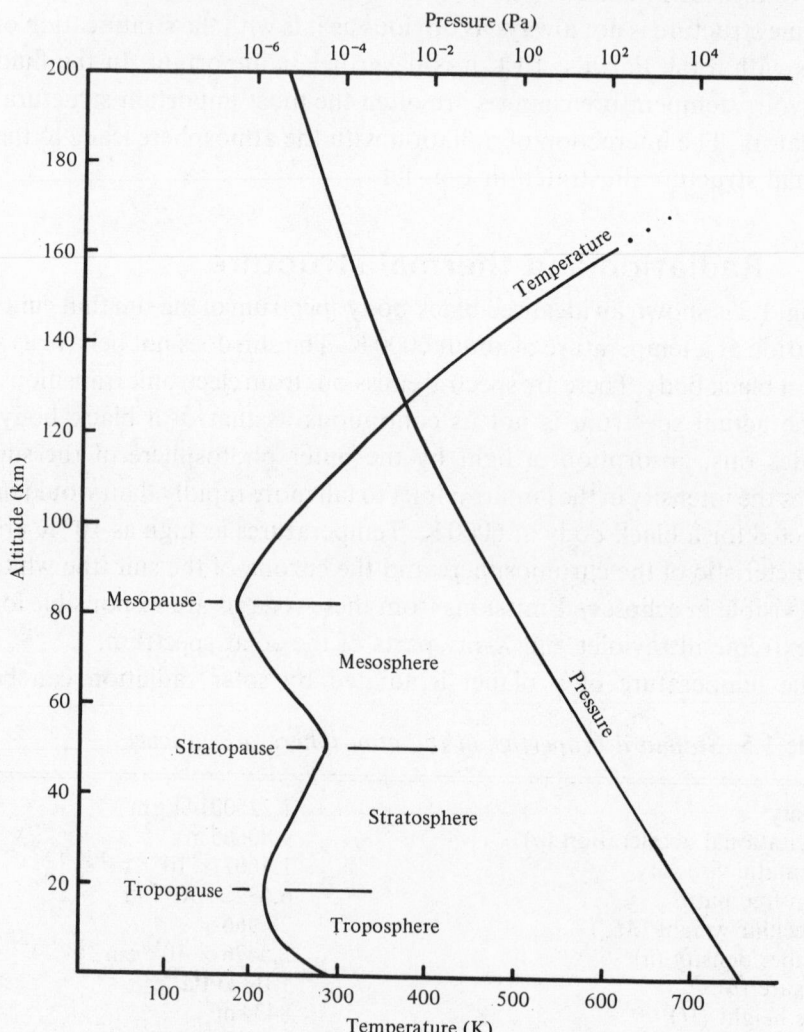

thought of as the area of the Earth's shadow, i.e. πr^2. Thus, the energy absorbed will be

$$R_a = I\pi r^2 (1 - A) \qquad (1.7)$$

The flux of outgoing radiation is given by the *Stefan–Boltzmann* Law, $I_0 = \sigma T^4$. Total energy radiated will be the product of this intensity and the area of the Earth, but this outgoing emission will take place from the entire surface so

$$R_o = 4\pi r^2 \sigma T^4 \qquad (1.8)$$

As R_a equals R_o (i.e. an assumption that the system is in steady state) an expression for temperature can be obtained:

$$T_e = [I(1 - A)/4\sigma]^{0.25} \qquad (1.9)$$

This is known as the effective temperature. The solar irradiance at the top

Fig. 1.2. A comparison between (a) the electromagnetic spectrum for black bodies at 6000 and 255 K and (b) the absorption spectrum of gases in the Earth's atmosphere. Note that the atmosphere is practically transparent to black body radiation emitted at temperatures typical of the sun.

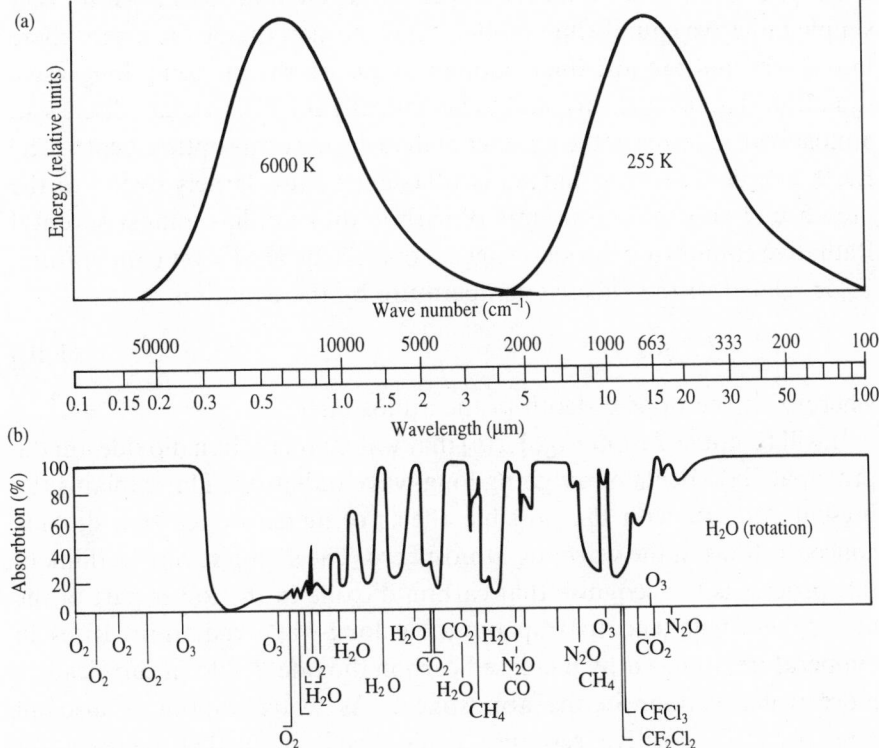

of the atmosphere is about $1.4 \times 10^3\,\mathrm{J\,m^{-2}\,s^{-1}}$ and the albedo 0.33. As the Stefan–Boltzmann constant is $5.67 \times 10^{-8}\,\mathrm{W\,m^{-2}\,K^{-4}}$, the equilibrium temperature of the Earth may be calculated as 254 K, which is far less than the average ground level temperature observed.

The black body spectrum of the Earth is shown in Fig. 1.2(a). Radiation from the Earth occurs at much longer wavelengths and is of much lower intensity than that of the sun. The lower portion shows the absorption spectrum of the Earth's atmosphere. The solar spectrum overlaps fairly well with a window in the absorption bands, while much of the long-wave emission from the Earth corresponds to a region of absorption in the atmosphere. The consequences of this are that much of the incoming radiation reaches the surface of the Earth, but the outgoing thermal radiation is largely re-radiated within the atmosphere and not all lost back into space. This 'greenhouse' effect results in higher temperatures than those calculated from Equation (1.9), which makes no allowance for the presence of an atmosphere.

In particular, the fact that the atmosphere is not transparent to the terrestrial long-wave radiation means that much of this radiation is absorbed in the lower part of the atmosphere. Because of this, the lower atmosphere is warmer than the upper parts. This can be shown by very simple radiative equilibrium models. Such models divide the atmosphere into layers that are just thick enough to absorb the outgoing long-wave radiation. These layers are said to be optically thick, so we can discuss an atmosphere in terms of the number of these layers or its optical depth. The Earth can be treated as having two layers: Venus, largely owing to the presence of enormous amounts of carbon dioxide, has almost seventy! Radiative equilibrium models suggest that the ground level temperature, T_g, is related to the effective temperature by the equation.

$$T_g^4 = (1 + \mathscr{T})T_e^4 \tag{1.10}$$

where $\mathscr{T}$ is the optical depth of the atmosphere.

It will be noticed from Fig. 1.2(b) that water and carbon dioxide are the principal absorbents of outgoing long-wave radiation. This explains the present concern with the possible effects of increasing carbon dioxide concentrations in the warming atmosphere. The simplest way to think of this process is to recognise that carbon dioxide in the lower part of the atmosphere will tend to trap outgoing long-wave radiation. Rises in temperature, induced by increased carbon dioxide, could in turn lead to more water vapour in the atmosphere. As water vapour is also an absorber, this positive response could result in further increases in temperature. The argument is more complicated than this because of the fact that carbon dioxide and water have many absorption bands of different strengths that play different roles at different altitudes.

Solving Equation (1.10) shows that, if the Earth had an atmosphere of two optical depths, then the equilibrium ground level temperature would be 334 K. This simple re-calculation of the radiation balance of the atmosphere overestimates the surface temperature. This is because it does not consider other processes that transport heat vertically, in particular convection, which is very important in lowering the temperature near the ground.

Convection occurs because warm air is lighter than cool air. As Warmer air rises it carries heat with it. The parcel of warm rising air must expand, as pressure decreases with altitude, and the work done will cause it to cool:

$$C_v \Delta T = - P \Delta V \tag{1.10}$$

where C_v is the molar heat capacity at constant volume and ΔV is the change in volume for the mole of air considered. The ideal gas equation is

$$PV = RT \tag{1.11}$$

which takes the differential form

$$P \, dV + V \, dP = R \, dt \tag{1.12}$$

This may be rearranged in the incremental form

$$- P \Delta V + R \Delta T = V \Delta P \tag{1.13}$$

Combining this equation with (1.10) and $C_p - C_v = R$ gives

$$C_p \Delta T = (C_v + R) \Delta T = - P \Delta V + R \Delta T = V \Delta P$$
$$= (RT/P) \Delta P \tag{1.14}$$

combining the equality $C_p \Delta T = (RT/P) \Delta P$ with Equation (1.4) gives the lapse rate

$$\Delta T / \Delta z = - M_m g / C_p \tag{1.15}$$

For the atmosphere of the Earth ($M_m = 0.028\,966 \text{ kg mol}^{-1}$, $g = 9.8065 \text{ m s}^{-1}$ and $C_p = 29.05 \text{ J mol}^{-1} \text{K}^{-1}$) this works out at -9.8 K km^{-1} for dry air, but the measured value of -6.5 K km^{-1} is due to the fact that the air is wet and, as it rises, condensation will release latent heat. If the temperature of the atmosphere falls much less slowly with height than the lapse rate (or even increases in temperature with height) then inversion conditions exist and the air is very stable with respect to vertical convective mixing. On the other hand if the temperature falls very rapidly with height (at a rate that is greater than the lapse rate) the atmosphere is unstable and convective mixing will be active.

Thus vertical transport processes and the absorption of re-emitted long-wave radiation explain the temperature profile of the lower part of the atmosphere or troposphere. However, we are still left with some surprising features in the vertical structure of atmospheric temperature. The foregoing paragraphs emphasised the transparency of the atmosphere to sunlight. While this is true for the visible part of the electromagnetic spectrum, the atmosphere is not transparent in the ultraviolet region. Although not much radiation is found at such short wavelengths, the small amount that is there has important consequences. The heights in the atmosphere at which various ultraviolet wavelengths are absorbed are shown in Fig. 3. Absorption by ozone at just over 40 km causes the upper atmosphere to warm up. This allows the average temperature of the poorly mixed stratosphere to be about 250 K.

Radiative transfer by carbon dioxide at stratospheric temperatures is

Fig 1.3. The depth of penetration of incoming short-wavelength solar radiation. There is a general increase in the penetration with increasing wavelength, except for the fact that hard X-rays may penetrate into the mesosphere.

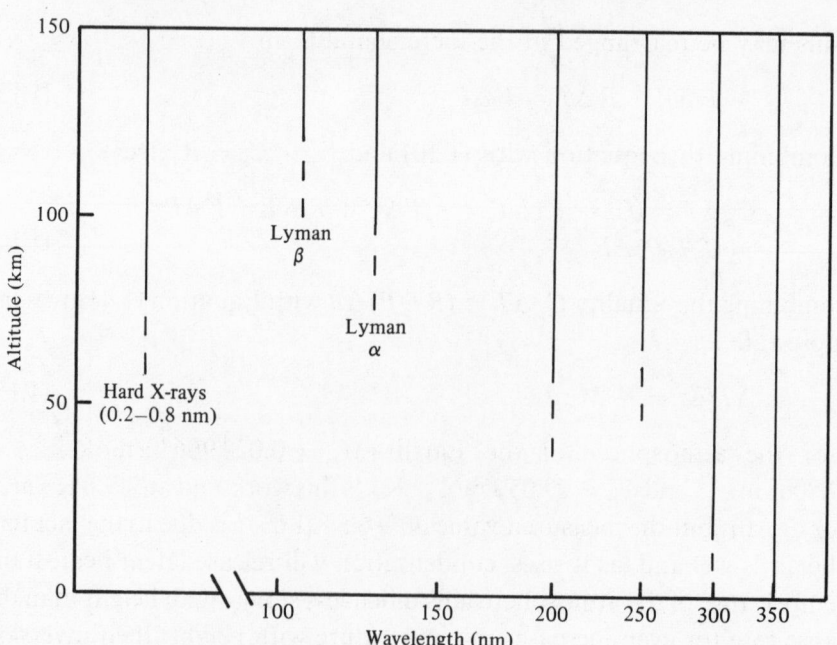

important in controlling the radiative balance of the stratosphere. Increases in carbon dioxide at these heights will enable the stratosphere to radiate more effectively and hence to become cooler. This means that increases in carbon dioxide, while causing a greenhouse warming in the lower atmosphere, will tend to make the stratosphere colder.

Another feature in Fig. 1.1 contradicts the notion of an atmosphere that declines in temperature with altitude. This is the rise in temperature found at extreme altitudes in the region known as the thermosphere. At such altitudes the atmosphere is very thin. Molecules are exposed to unattenuated solar radiation of extremely short wavelength (high energy). High energy radiation arises from emissions of the outer region of the sun. At wavelengths below 50 nm, the effective emission temperature exceeds 10 000 K. High-energy solar photons of such short wavelength transfer energy to gas molecules to give them high translational velocities, i.e. high temperatures. The energies may be large enough to dissociate O_2 and N_2. The temperatures found in the thermosphere undergo wide variation depending on the state of the sun. In times of solar disturbances the output of hard photons (i.e. high energy) is much enhanced. The temperature of the thermosphere is further increased by another mechanism. While temperature is normally defined in terms of translational energy, absorption and emission of radiation occur through vibrational and rotational changes. In the upper atmosphere there are relatively few collisions between molecules, so the opportunities to exchange translational, vibrational and rotational energy are infrequent, therefore the cooling of the thermosphere is inefficient. Thermospheric temperature increases with height so it is also stable against convection. Heat can only be lost by slow diffusive processes. The result is a region of the atmosphere where temperature gradients are extremely high.

1.5   General circulation

It has been long been known that, in many parts of the world, there are certain winds that prevail for much of the year. In the days of sailing vessels they were given fanciful names such as the *trade winds* or the *doldrums*. These winds are part of the general circulation of the atmosphere, which is driven by the input of solar energy and modified by the rotation of the Earth. Understanding the general circulation is a fundamental problem of meteorology and a complete description of the circulation still poses some problems. Here we shall consider only a few of the simpler features that are important in redistributing trace components through the atmosphere.

Wind directions in the summer and winter months are shown as kinds of mariner's charts in Fig. 1.4(*a*) and (*b*). In Fig. 1.4(*c*) are shown regions of sea haze and superimposed on these continental areas are shown the calculated distributions of industrial and agricultural sources of particle pollution, but much of the haze is probably not of human origin.

Fig. 1.4. Winter (*a*) and summer (*b*) surface winds compared with the principal areas of maritime haze (*c*) taken from Arrhénius (1959). The dotted lines in (*a*) and (*b*) represent the average position of the ITCZ. There is also given in (*c*) a notional representation of the agricultural and industrial source strengths. The small dots are equivalent to the direct industrial emission of a $Tg\,a^{-1}$ of particulate material, while the dashes symbolise a similar quantity from agricultural sources.

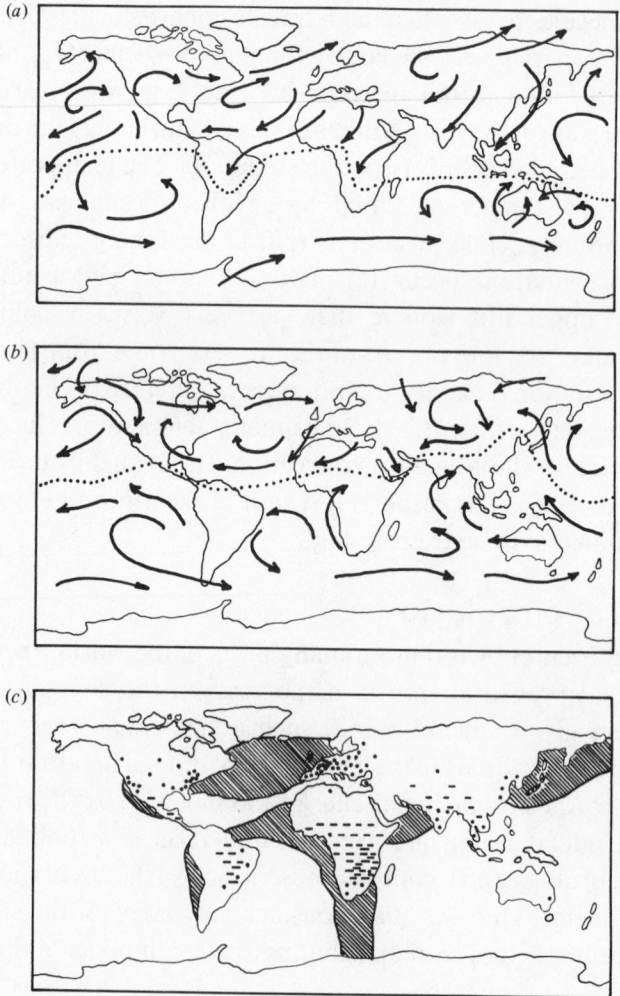

Pollutant particles are carried by strong summer westerlies from North
America out across the Atlantic Ocean. This shows up clearly in Fig.
1.4(*c*) as an area of summer haze in the North Atlantic. In the winter (see
Fig. 1.4(*b*)) there is a strong flow down across the Arabian Peninsular
carrying with it pollution and wind-blown dusts from the deserts. This
appears in Fig. 1.4(*c*) as an area of winter sea haze. The winds that cross
the Sahara (Harmattan winds) and blow dust to the east are present in
both Fig. 1.4(*a*) and (*b*) and the spreading of Sahelian dust across the
Atlantic to Bermuda is a well-known phenomenon.

These diagrams do not show the vertical structure of the atmosphere,
yet an account of vertical movement is essential in any description of
general circulation. In Fig. 1.5 are shown both the vertical and the
horizontal movements of the air. The circulation is divided into three cells
by vertical movements of air masses. Southerly and northerly air flows of
air meet at ground level near the equator in a region known as the
intertropical convergence zone (ITCZ). Here warm tropical air rises and,
in contrast, at the poles descending air masses are found. Outside the
tropics, circulation is dominated by large rotational air motions or
cyclones. These develop at the polar front where temperature contrasts
between polar and tropical air masses are most pronounced. Occasionally
outbreaks of polar air can penetrate into the tropics, ensuring hemispheric

Fig. 1.5. A simple representation of global circulation (from Smith (1982).

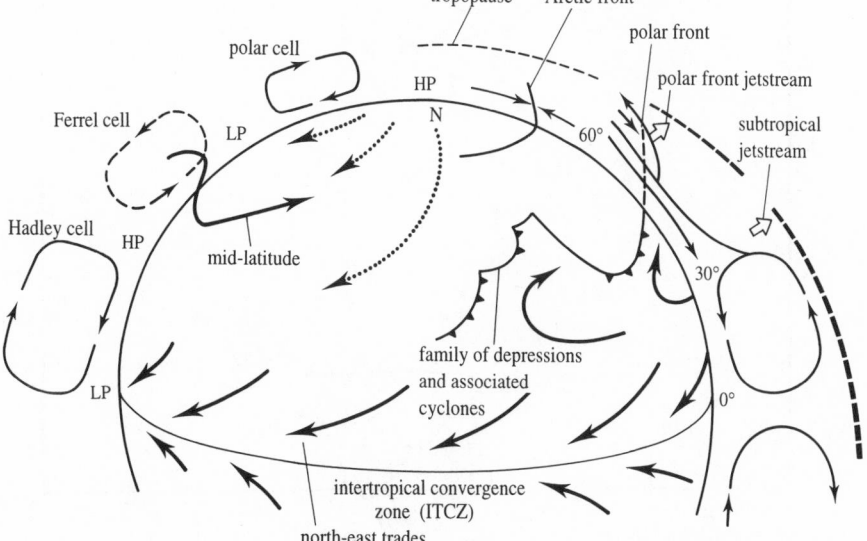

mixing. However it is difficult for gases to mix across the equator, as suggested by the existence of the ITCZ, so it is not uncommon for trace gases in the two hemispheres to have different concentrations. This is illustrated for a few trace components in Fig. 1.6. Here the concentrations of a number of gases that are generated by anthropogenic activities in the Northern Hemisphere are shown as a function of latitude. Surface concentrations tend to be higher in the northern hemisphere, although the change tends to be more pronounced with a short-lived gas such as carbon monoxide in comparison with a long lived freon. At higher altitudes the sharp transition across the ITCZ is no longer apparent.

Another illustration of the way in which vertical motions in the atmosphere affect composition is shown by the elevated concentrations of sea salt that have been found at great distances from the sea in the snows of central Antarctica. These have been attributed to the upward transport of air rich in sea-salt aerosol at the polar front followed by deposition onto

Fig. 1.6. Latitudinal variation of the concentration of two pollutant gases. At ground level the effect of the ITCZ on the concentration of carbon monoxide can be very marked. Smoothed values for Freon-11 concentrations suggest that slightly lower concentrations are to be found in the Southern Hemisphere.(Data taken from Junge and Seiler (1971) and Hunter-Smith R.J. et al. (1983).

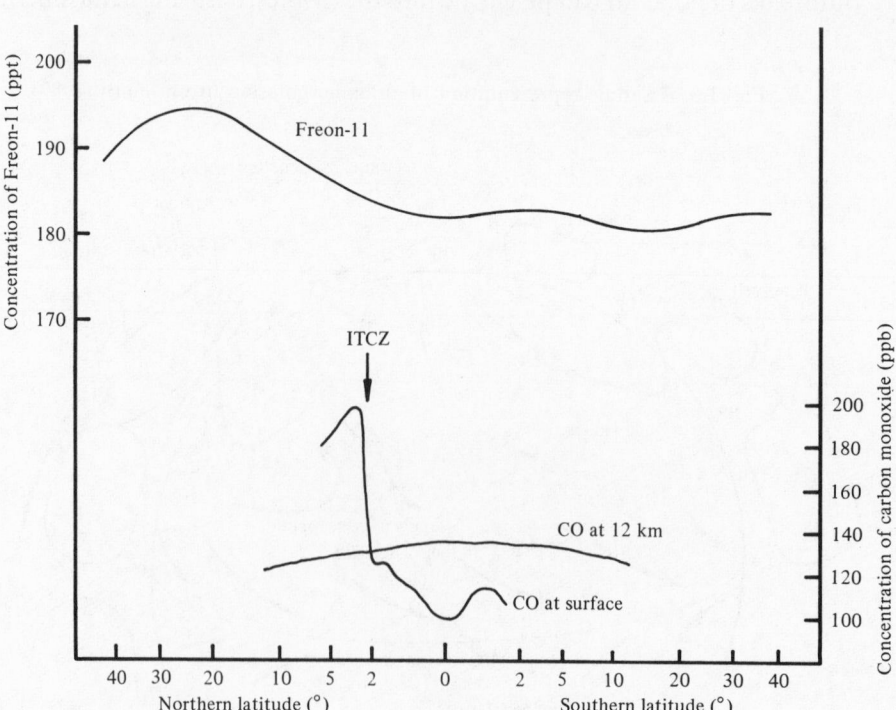

the snow after being brought to low altitudes by the air that descends over the poles.

Air in the stratosphere is not restricted by the cellular flow evident in the troposphere. This means that inter-hemispheric transport is easier in the stratosphere. Tropospheric air can move into the stratosphere relatively easily at points where the tropopause breaks down, especially near the sub-tropical jet streams. An example of tropospheric air leaking into the stratosphere may be seen in Fig. 1.7 that shows the concentrations of ^{85}Kr in the atmosphere. Radioactive debris from large nuclear explosions, dust volcanoes and possibly even materials within thunderstorms can break through the tropopause and fallout and other material can enter the stratosphere. Once there, air movements and long residence times mean that they are readily distributed over the entire globe.

1.6   Water

Water is one of the most important components of the atmosphere, although it comprises less than 2% of the total volume. Almost a quarter of the energy in the solar radiation that falls on the Earth is used to evaporate water. The enormous amount of energy used in evaporation is

Fig. 1.7. Global distribution of ^{85}Kr with altitude. This gas mainly originates from nuclear reactors and shows the typical distribution expected for a long-lived radioactive gas with a source at ground level. Note the low altitudes of the distribution over the poles and the vertical excursions that seem to correspond to breaks in the tropopause. The gas would, in general, cross the tropopause at such locations. Nuclear or volcanic explosions can push materials through the tropopause directly. Krypton-85 data from Telegadas and Ferber (1975).

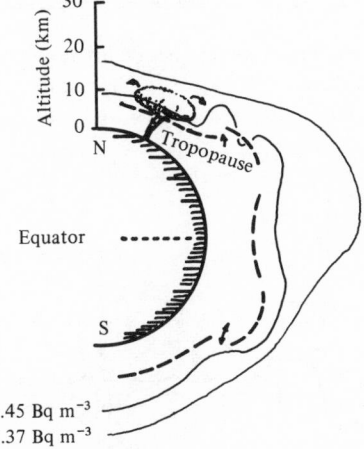

partly because water has an extremely high latent heat of vaporisation. This is only one of several unique properties of water, many of which are the result of hydrogen bonding.

The cycling of water through the atmosphere is shown in Fig 1.8. Like all simple depictions of global cycles, it neglects geographical variations. Table 1.6 shows the water balance for the various continents and we see that South America, which contains most humid tropical rain forests, has large fluxes of water. Australia, with its enormous central desert area has much lower fluxes.

The amount of water in the atmosphere is less than a thousandth of a percent of all the water available to the hydrological cycle yet its effects on

Table 1.6. *Water balance of the continents expressed in millimetres per annum*

	Precipitation	Evaporation	Run-off
South America	1350	860	490
North America	670	400	270
Africa	670	510	160
Asia	610	390	220
Europe	600	360	240
Australasia	470	410	60

Table 1.7. *Saturation vapour pressure of pure water*

Temperature (°C)	0	5	10	15	20	25
Pressure (Pa)	610	872	1228	1705	2338	3167
Water content ($g\,m^{-3}$)	4.847	6.797	9.399	12.83	17.30	23.05

Fig. 1.8. A global cycle for water. The numbers beside arrows give the flux of water from one reservoir to another in $10^{18}\,kg\,a^{-1}$. The numbers within reservoirs give their water content in $10^{18}\,kg$.

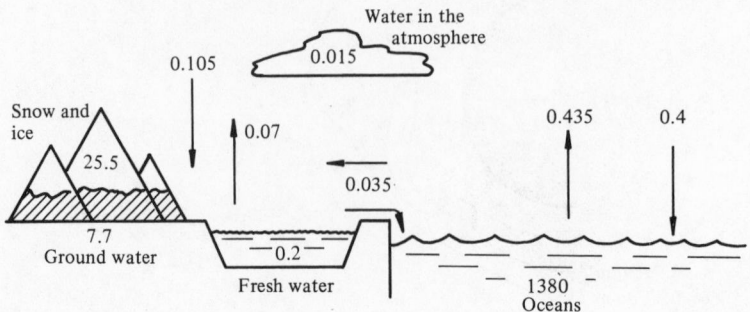

the atmosphere are enormous. We have already seen how it lowers the lapse rate in the troposphere. Table 1.1, which lists the components of the atmosphere, neglects water vapour because its amount is extremely variable. One reason is the sharp dependence of vapour pressure and water content of air on temperature as shown in Table 1.7. Not only does this lead us to expect to find less water in air near the poles compared with that from tropical regions, but it also means that there is a decline in water content with altitude.

Furthermore, the atmosphere is not saturated with water vapour and the expression of the degree of undersaturation, or relative humidity, is a familiar concept. The relative humidity simply represents the ratio of the measured water vapour pressure to the saturation vapour pressure at the temperature in question. The relative humidity is usually expressed as a percentage. It is important to remember that the water in the atmosphere is present in the solid and liquid phases too and that these phases are responsible for so many weather phenomena, e.g. clouds, fogs, rain and snow.

Evaporation from the oceans is the main source of water and the main sink is precipitation (rainfall etc.). The average residence time of water in the atmosphere is about ten days, although this varies with latitude. Residence times of as long as fifteen days are found at extremely high latitudes and in middle latitudes it can be less than seven days. If we recollect the excellent solvent properties of water then it will be evident that many atmospheric trace components will dissolve in liquid water in the atmosphere and this will mean that they will also tend have a residence time of around ten days.

1.7   Further reading

Smith, D.G. (1982) *Cambridge Encyclopaedia of Earth Sciences*, CUP, Cambridge.

2

○ ○

The natural components of the air

Chapter 1 dealt with the bulk composition of the atmosphere and although this presented a great challenge to early chemists, atmospheric chemistry has increasingly come to be the study of the variable-trace components of the atmosphere. These, despite their low concentration, exert a profound influence on the general chemistry of the atmosphere.

The scientist James Lovelock has drawn attention to the fact that the atmosphere is very far from chemical equilibrium in his Gaia Hypothesis. Oxygen, for example, is an extremely reactive element and should react with a whole range of materials at the Earth's surface. Equilibrium considerations would also suggest that oxygen should combine with the nitrogen in the atmosphere. We do not find significant levels of oxygen on either Mars or Venus. Yet on the Earth the atmosphere is considerably out of equilibrium, so we can presume that some very unusual processes are at work. On the Earth we know that many large departures from the expected equilibrium state are due to the activities of living organisms.

2.1   Biological sources

The effect of living organisms on the composition of the atmosphere is clear. The fact that many familiar organisms respire and release carbon dioxide and that this results in oxygen via photosynthesis is common knowledge. The quantities involved in these biological processes are enormous. The 10^{18} kg of oxygen present in the Earth's atmosphere is virtually all the result of photosynthesis. The high concentration is maintained despite continual removal in oxidative processes. In simple terms the processes of respiration and photosynthesis may be written symbolically as

$$\text{`CH}_2\text{O'} + \text{O}_2 \;\underset{\text{photosynthesis}}{\overset{\text{respiration}}{\rightleftharpoons}}\; \text{CO}_2 + \text{H}_2\text{O} \qquad \text{(R2.1)}$$

The reservoir of organic carbon is small compared with the amounts of available oxygen in the atmosphere. This means that in the long term it is not organic carbon that would control the atmospheric oxygen. On these longer time-scales the concentration of oxygen in the atmosphere is controlled by the oxidation of weathered minerals containing oxidisable elements such as iron. However the availability and the rapid reactions of organic material mean that at present it is carbon that exerts a dominant control on the oxygen–carbon dioxide balance of the atmosphere.

If we consider gases of biological origin, then some reasons for variability in concentration spring to mind. The presence of large numbers of photosynthesising organisms will obviously enhance the oxygen concentration in a given area, although the oxygen concentration of the atmosphere is so high that this effect will be very slight. On the other hand, at night when there is no light for photosynthesis, respiratory carbon dioxide from the same organisms will raise the level of this gas in the atmosphere. As carbon dioxide concentrations are low, the effect of biological activity is quite marked. The air in dense forests can be enriched in carbon dioxide at night by perhaps 10 ppm relative to background levels. During the day, photosynthesis can drive this some 25 ppm lower than the background level. In soil air, where ventilation is obviously restricted the concentrations of carbon dioxide can be as high as 0.5% (5000 ppm).

Compounds of carbon

Carbon dioxide and oxygen are major components of the atmosphere rather than the trace gases that are the subject of this chapter. It often seems that the more exotic gases (i.e. those considerably out of thermodynamic equilibrium) are produced by micro-organisms. Some of these gases are also produced by more familiar higher plants and animals, but they may be produced through the presence of associated micro-organisms. In animals this is most frequently from micro-organisms in the gut.

Methane, the major hydrocarbon in the atmosphere, is produced in marshes and paddy fields through the microbial degradation of organic matter that may be written as

$$\text{CO}_2 + 4\text{H}_2 \rightarrow \text{CH}_4 + 2\text{H}_2\text{O} \qquad \text{(R2.2)}$$

The hydrogen for this reaction is usually derived from degradation of alcohols by non-methanogenic bacteria in the same environment. The intestines of termites and cows also provide food materials and a suitable environment within which micro-organisms can produce methane.

Larger hydrocarbon molecules can also be produced through biological activity, although the release rates of ethane and propane from soils are less than a percent of that for methane. Ethene (ethylene) on the other hand is more abundant than these higher alkanes. It is produced by growing plants and acts as a growth regulator, which is critical in controlling senescence, fruit ripening and other metabolic activities. Increases in the production of ethene are also associated with stress in plants, and even small quantities of toxins or slight injuries can initiate ethene production.

Forests emit large amounts of complex hydrocarbons. Isoprene $(CH_2=C(CH_3)CH=CH_2)$ appears to come from non-coniferous forests of oak, poplar, sycamore, willow etc. Pine forests are likely to be associated with monoterpenes, such as α-pinene and β-pinene, with a lesser quantity of other terpenes and oxygenated compounds, such as 6,6-dimethylbicyclo[3.1.1]heptane-2-one. The total amount of these plant-derived hydrocarbons released on a global basis is estimated to be probably in excess of $800 \, Tg \, a^{-1}$. Deciduous forests and grasslands are also sources of lesser quantities of simpler alkanes and alkenes. At extremely low concentrations, we are aware of the odours of flowers and

Table 2.1. *Concentration of hydrocarbons in the remote continental boundary layer*

	Concentration (ppb)
Methane	1600
Ethane	1–6
Ethene	0.2–3.2
Ethyne (acetylene)	0.3–2.4
Propane	0.1–2.7
Propene	0.2–0.8
n-Butane	0.1–1.9
1-Butane	0.1–0.7
n-Pentane	0.1–0.7
Benzene	0.2–1.2
Toluene	0.1–1.0
Isoprene	0.6-2.3
α-Pinene	0.033–1.5
β-Pinene	0.1–0.43

fruits as perceptible emissions from plants. Table 2.1 gives typical values for the hydrocarbon concentrations in continental air. It is important to notice how low the concentrations are, given the enormous amounts thought to be emitted into the atmosphere. The most likely explanation for this is rapid loss of these compounds through oxidation.

Some evidence in support of this may come from the high frequency of hazes over forests. Forest haze can arise from air pollution transported over long distances, but natural hazes can be formed through oxidation of terpenes and isoprene in air. Such hazes are frequent under warm conditions and it is at these times that essential oils can volatilise from trees and photochemically induced oxidation occurs most readily. In sunny weather much of the terpenoid carbon can be oxidised to particulate matter in a few hours.

Animals also emit a range of trace substances besides methane. The largest quantities are as bio-effluent hydrocarbons, aldehydes and reduced sulphur compounds from the gut or in expired air. Much smaller quantities of volatile compounds such as pheromones are deliberately released as olfactory signals. Some aphids, for example, emit *trans-β-farnesene* when under attack by predators. This warns other aphids so that they can escape. Many ants use ketones when under attack (even simple ones such as 3-octanone) to stimulate aggression. Other organic compounds are utilised as sexual attractants. The concentrations of these are often so low that to sense their presence in the atmosphere, animals may need to have receptors capable of detecting concentrations of less than 100 molecules cm^{-3}. In this way quite modest releases of pheromone ($100\,pg\,s^{-1}$) may be detected up to 100 m away. Pheromones are widely spread in the animal kingdom. It has even been suggested that human interest in musky perfumes is related in an evolutionary way to earlier sex attractants.

Compounds of nitrogen
Despite these examples of specific releases of trace substances by animals, the trace gas emissions most often encountered are a result of microbial activity. Ammonia emissions in the atmosphere can be well correlated with the presence of animal urine, but much of the gas is generated by micro-organisms that are responsible for degradation of nitrogen compounds within the urine.

Nitrogen present as urea in urine can readily be hydrolysed to ammonia and carbon dioxide according to the equation

$$NH_2CONH_2 + H_2O \rightarrow 2NH_3 + CO_2 \qquad \text{(R2.3)}$$

If the soil in which this hydrolysis occurs is alkaline, gaseous ammonia can be released. If it is more acidic, the ammonia will be rendered less volatile through formation of the ammonium ion (NH_4^+). Nitrogen in living matter is often bound within proteins built from amino acids. Amino acids in urine can also be broken down to ammonia, e.g. glycine can be oxidised in the following way:

$$CH_2NH_2COOH + \tfrac{3}{2}O_2 \rightarrow 2CO_2 + H_2O + NH_3 \qquad (R2.4)$$

Plants can absorb soil ammonia or ammonium directly. Some micro-organisms, such as *Nitrosomonas*, can oxidise ammonia thus using it as an energy source in the same way as that in which other organisms use reduced carbon compounds (e.g. carbohydrates as in (R2.1)):

$$NH_3 + \tfrac{3}{2}O_2 \rightarrow H^+ + NO_2^- + H_2O \qquad (R2.5)$$

Nitrobacteria can oxidise this nitrite to nitrate:

$$NO_2^- + \tfrac{1}{2}O_2 \rightarrow NO_3^- \qquad (R2.6)$$

They may also produce nitrous oxide (N_2O). The numerous micro-organisms responsible for the reverse process, denitrification, can also produce N_2O. Under these processes soil nitrate is reduced by aerobic bacteria, which can use nitrate as a hydrogen acceptor when oxygen concentrations become limiting. The energy provided by this nitrate respiration can be nearly an order of magnitude greater than that provided by the oxidation of ammonia and almost as much as that provided by aerobic respiration. Nitrate respiration can be represented by the equations

$$C_6H_{12}O_6 + 6NO_3^- \rightarrow 6CO_2 + 3H_2O + 6OH^- + 3N_2O \quad (R2.7)$$

$$5C_6H_{12}O_6 + 24NO_3^- \rightarrow 30CO_2 + 18H_2O + 24OH^- + 12N_2 \qquad (R2.8)$$

Both these denitrification reactions yield gaseous emissions. Reaction (R2.7) is an obvious source of nitrous oxide in the atmosphere. Such reduction reactions of nitrate may also give rise to nitric oxide sometimes.

Compounds containing sulphur

Mass balance calculations have long suggested that biological emissions of sulphur need to be considered in order to account for the large flux of sulphur compounds into the atmosphere. Initially it was thought that hydrogen sulphide, a product of the reduction of sulphate in anoxic sediments, would be a major biological source of gaseous sulphur in the atmosphere. However, although large quantities of hydrogen sulphide are produced in these sediments, much of it ultimately becomes bound within minerals, such as pyrites, instead of being lost to the atmosphere. Hydrogen sulphide is also susceptible to oxidation in aqueous systems.

Since the 1970s it has become apparent that organosulphides produced by micro-organisms make a most significant contribution to the atmospheric sulphur burden. In particular the compound dimethylsulphide (DMS, $(CH_3)_2S$) is an important component of both marine and soil emissions. It is produced by marine phytoplankton, such as *Phaeocystis pouchetii*, in the upper layers of the ocean and represents the major flux of reduced sulphur to the marine atmosphere. This DMS is produced by the hydrolysis of the organosulphur species β-dimethylsulphoniopropionate (DMSP, $(CH_3)_2S^+CH_2CH_2COO^-$) to DMS and acrylic acid

$$(CH_3)_2S^+CH_2CH_2COO^- \rightarrow (CH_3)_2S + CH_2{=}CHCOOH$$
$$(R2.9)$$

Maximum concentrations of DMS in the oceans are found at, or a few metres below, the surface. The concentrations can be as high as several micrograms per litre, but they fall off rapidly with depth, because phytoplanktonic species are not active in the absence of light. The precursor DMSP is often present at concentrations an order of magnitude higher than that of DMS. As DMS is rather volatile and insoluble it is rapidly lost from seawater to the atmosphere.

A number of other volatile reduced sulphur compounds have been found in the ocean. These, methylmercaptan, dimethyl disulphide, carbonyl sulphide and carbon disulphide also contribute to the flux of sulphur to the atmosphere. Carbonyl sulphide can be produced by the oxidation of carbon disulphide in the atmosphere or its hydrolysis in seawater:

$$CS_2 + H_2O \rightarrow COS + H_2S \qquad (R2.10)$$

The oceans have a high sulphate content and so are a good place for micro-organisms to generate sulphur compounds. However similar reduced sulphur compounds can also be produced in those soils in which sulphur is abundant. The production of reduced sulphur compounds by soils is most effective at high temperature, so it is likely that tropical soils are particularly important, with wetlands being a major source despite their limited area.

Compounds containing other elements

Halogenated organic compounds are well known in the atmosphere, although these have often been regarded as a product of human activity. As halogen compounds in the atmosphere have received more attention, it has become apparent that there are many biological sources. Methyl chloride is the most abundant halocarbon in the atmosphere and probably the most important natural source of stratospheric chlorine. Much of the methyl chloride comes from the ocean through biological processes that are still poorly understood. On land microbiological processes and the combustion of biomass yield further methyl chloride. Methyl bromide and a range of mixed bromochloroalkanes have been identified in seawater. Methyl iodide is released from the oceans in considerable quantity, although the gas is photochemically reactive which limits its lifetime and concentration in the atmosphere.

Phosphorus compounds are usually regarded as too involatile to have an important trace gas chemistry in the atmosphere. However, 'will'o'wisps', the luminous phenomenon found over marshy ground, have sometimes been attributed to the ignition of methane by traces of phosphane (PH_3) or diphosphane (P_2H_4). Some bacteria are able to produce these phosphorus-containing gases, together with methane, under the strongly reducing conditions found in the intestines of ruminants.

Many heavy metals, although inessential, and frequently toxic, to organisms are metabolised to volatile compounds. The concentrations under natural conditions must be fairly low, except in regions of heavy mineralisation or anthropogenic pollution. The most general process is methylation. It is now fairly well understood in the case of mercury, for which the compound methyl cobalamin (which is a close relative of vitamin B-12, and is often represented by the formula CH_3CoB_{12}) usually acts as the source of methyl groups in the methylation:

$$CH_3CoB_{12} + Hg^{2+} + H_2O \rightarrow H_2OCoB_{12}^+ + CH_3Hg^+ \quad (R2.11)$$

Further methylation yields the volatile dimethyl mercury that can escape to the atmosphere. Alternatively the formation of methyl mercury chloride, which has an appreciable vapour pressure, gives detectable quantities of mercury in the atmosphere. However, the main source of atmospheric mercury is microbial reduction and elemental volatilisation. Lead, thallium, tin and cadmium could also be methylated, but alkylated cadmium compounds are not likely to be important as they are unstable in aqueous solution. The emission of pollen grains and leaf fragments also provides a biological source of metallic elements to the atmosphere.

The non-metals arsenic, selenium and tellurium can also be alkylated. The release of arsenic by moulds has been known since the nineteenth century. *Scopulariopsis brevicaulis* will produce trimethyl arsine $((CH_3)_3As)$ from media containing arsenious oxide. Bacteria and moulds are also able to produce dimethyl arsine. In the atmosphere these arsines are likely to be oxidised to cacodylic acid $((CH_3)_2AsO_2H)$. Antimony seems less to be readily methylated than arsenic. Biological production of methylated selenium and tellurium compounds by moulds is well known. Measurements in the atmospheres of riverine environments have established the presence of dimethylselenide $((CH_3)_2Se)$, dimethyldiselenide $((CH_3)_2Se_2)$ and possibly dimethylselenone $((CH_3)_2SeO)$ in nanogram amounts per cubic metre. Organoselenium and organotellurium compounds are ventilated by animals fed selenite or tellurite salts. Selenium compounds in the marine atmosphere appear distributed between the vapour and solid phase. This may suggest the presence of both organic and inorganic selenium compounds.

2.2   Geochemical sources

The most important geochemical source of trace gases at the surface of the Earth are volcanoes. These are massive sources, but ones that are highly localised and extremely variable over time. When Mount Etna is erupting it can emit sulphur dioxide at a greater rate than all the industries of Europe put together, but at other times it emits only tiny amounts. When eruptions are particularly violent, volcanic emissions can be injected into the stratosphere, although volcanic emissions include the much more gentle continuous flux of gases from fumaroles.

Large quantities of carbon dioxide can be released by volcanoes, but it is probably the sulphur gases that represent the most significant contribution that volcanoes make to the budget of trace components in the atmosphere. The volcanic emissions take the form of sulphur dioxide or hydrogen sulphide and to a lesser extent carbonyl sulphide or even carbon disulphide.

Halogens are also emitted in large quantities, usually as the hydrohalic acids, but small amounts appear in volcanic gases as methyl chloride, bromide and iodide. It is not entirely clear how organohalides form in volcanic emissions, as they would not be predicted on the grounds of simple equilibria in volcanic vents. Some have argued that they are generated through reactions of volcanic gases with vegetation. This mechanism would also explain the presence of some complex hydrocarbon compounds in plumes. Estimated fluxes of volcanogenic gases to the atmosphere are listed in Table 2.2. However, there is much concern about the accuracy of these estimates given the difficulty of summing up the emissions across the range of very different volcanoes that occur worldwide.

Emissions from volcanoes can cause great damage. In major eruptions, the eruptive activity is usually the greatest hazard although sometimes clouds of gases can asphyxiate people before they can escape. Continuous volcanic emissions of a gentler kind can have a more subtle effect. The Poas volcano of Costa Rica emits large quantities of sulphur and chlorine compounds that injure both human health and crops in the important agricultural region that surrounds the volcano.

The eruption of volcanoes, like the action of wind at the surface of the Earth is the source of large amounts of particulate material. These sources of dust, together with the oceans as a source of sea-salt particles, are large and important enough to warrant a more complete discussion, in Chapter 4.

Rocks themselves can be the source of small amounts of gas. These emanations are the dominant sources of the noble gases: helium, argon and radon. Argon has accumulated over geological time from the radioactive decay of ^{40}K on the Earth. Helium and radon also arise as decay products, but they do not accumulate, because there are effective sinks. Helium is largely produced from the decay of ^{238}U and ^{232}Th, but is so light that it can escape from the exosphere at the very top of the Earth's atmosphere. The gas ^{222}Rn is one of the products of the uranium decay series. It has a half life of 3.8 days, which means that its presence in the atmosphere is relatively ephemeral. However, the fact that it has this short half life and arises from radioactive decay over the continents means that it is a good marker for continental air.

Table 2.2. *Annual emissions of volcanic gases. The global flux is taken in terms of the weight of the species*

Gas	Global flux ($Tg\,a^{-1}$)
Sulphur dioxide	10
Carbonyl sulphide	0.02
Hydrochloric acid	8.0
Hydrofluoric acid	0.4
Hydrobromic acid	0.08
Methane	0.34
Boron	0.3
Hydrogen	0.2
Carbon monoxide	0.02
Methyl chloride	0.001
Mercury	0.0001

Over the past decade or so it has become apparent that many heavy metals such as lead, mercury and cadmium, which have low melting points, are present in the atmosphere at concentrations that are considerably more than would be expected from weathering of crustal materials. There has been some speculation that these elements may evaporate directly from rocks at the Earth's surface. However, it seems that, at ambient temperature, this is not likely to be an important source, but evaporation might be significant in thermal areas or from volcanoes.

Besides this there is also a small input from material of extra-terrestrial origin and from reactions induced by incoming cosmic radiation. Despite the small input flux, meteoritic sodium, and to a lesser extent other metals, has observable effects on the behaviour of the upper atmosphere. We will examine these further in Chapter 8.

The oceans can also represent a significant source of gases, although many of these originate through biological activities within the oceans. We saw this in the case of some volatile sulphur compounds such as dimethylsulphide. Because so little is understood about the source of many of these compounds, they are often treated as if they come from the ocean as a whole. The ocean is often a source because the gases in question are present at supersaturated concentrations in the sea. These high concentrations mean that they readily diffuse across the air–sea interface into the atmosphere. Table 2.3 lists some gases for which the ocean is a source. Boron loss from the oceans represents a major source for this element in the atmosphere. This is assumed to be vapour as boric acid (H_3BO_3), although a small fraction of the atmospheric burden is present in the particulate phase. Volcanoes and human activities represent further sources of atmospheric boron. Table 2.3 also shows that a range of other non-volatile elements, such as the metals, can be released by the sea.

Next we consider as a source of trace gases forest or savannah fires. When considering fires as natural sources of trace gases, it can be hard to decide which fires are natural and which were either deliberately lit or indirect results of human activities. Forest fires have been increasingly studied in the last decade as significant sources of smoke and gaseous pollutants. They produce large quantities of unburnt and pyrolised organic compounds, methyl chloride, carbon monoxide and nitrogen oxides. They also produce less significant quantities of sulphur compounds such as sulphur dioxide and carbonyl sulphide.

It has not been easy to estimate the size of the global emissions because both the number and size of fires and the emission factors for individual compounds have to be established. Emission factors are required in order to estimate the amount of pollutant that arises from burning a given amount of fuel or combustible substance. These have been determined from studies, often in the laboratory, of the release of gas from burning and smouldering vegetation. Fires represent an important source of some compounds and probably contribute about half the naturally emitted carbon monoxide, hydrogen cyanide, methyl chloride and elemental carbon to the atmospheric inventory (see Table 2.4). Perhaps a little more surprising is the fact that they also make a noticeable contribution to the flux of metallic elements to the global atmosphere.

2.3   Atmospheric sources

The large fires mentioned above not only add trace gases and particles to the atmosphere, but also through the presence of these materials perturb the chemistry of the atmosphere. Plumes from forest fires are often

Table 2.3. *The flux of a number of trace gases and metals from the oceans. The global flux is taken in terms of the weight of the species*

	Global Flux ($Tg\,a^{-1}$)
Carbon monoxide	100
Dimethyl sulphide	30
Hydrogen sulphide	15
Methane	10
Nitrous oxide	6
Methyl chloride	5
Methyl bromide	0.3
Methyl iodide	0.5
Hydrogen	4
Boron (as H_3BO_3)	1–5
Ethane	1
Carbonyl sulphide	0.8
Chloroform	0.7
Carbon disulphide	0.3
Antimony	0.00005
Arsenic	0.0023
Cadmium	0.00005
Cobalt	0.00008
Copper	0.00039
Mercury	0.00077
Manganese	0.0015
Nickel	0.00012
Lead	0.00024
Selenium	0.0047
Vanadium	0.00016

associated with the production of elevated ozone concentrations. Reactions that take place in the atmosphere represent a further source of many trace gases. One process suggested very early was the possibility that lightning would fix atmospheric nitrogen. There has subsequently been a long debate over the size of this source and its role in adding all-important nitrogen compounds to soils. The reaction proceeds when nitrogen and oxygen, in air, are heated to high temperatures by the shock waves from a lightning stroke:

$$N_2 + O_2 \rightarrow 2NO$$

The yield of this process is dependent on both the temperature and the rate of cooling. If the reacted gas mixture is cooled slowly there is time for the reactants to be reformed, so the yield of nitric oxide is low. In lightning, temperatures can change very rapidly. Temperatures as high as 4000 K may be reached in the discharge channel, where the oxidation of nitrogen proceeds rapidly.

It is difficult to assess the exact size of this source, but it is likely that

Table 2.4. *The estimated global emissions from biomass-burning (much is adapted from Crutzen and Andreae, (1990) and Nriagu (1989)). Global flux is taken in terms of the weight of the species*

	Global flux $(Tg\,a^{-1})$
Carbon dioxide	1600–4100
Carbon monoxide	120–510
Methane	11–53
Ethene	3
Hydrogen	5–16
Methyl chloride	0.5–2
Nitrogen oxides	2.1–5.5
Hydrogen cyanide	0.4–1.9
Methyl cyanide	0.14–0.8
Sulphur dioxide	1–4
Carbonyl sulphide	0.04–0.2
Arsenic	0.00019
Cadmiun	0.00011
Cobalt	0.00031
Chromium	0.00009
Copper	0.0038
Mananese	0.023
Nickel	0.0023
Lead	0.0018
Vanadium	0.0018

lightning plays an important part in the fixation of nitrogen. There is general agreement that about 4×10^{26} molecules of nitric oxide will be produced in a single cloud-to-ground stroke. There are many such flashes over the globe all the time, which has led to estimates of a production rate of $8\,Tg(N)\,a^{-1}$. Nitrogen will also be fixed by cloud-to-cloud strokes, but the frequency and production rates of these are not as well known. At high concentrations, the nitric oxide produced by lightning will be rapidly oxidised to nitrogen dioxide.

$$2NO + O_2 \rightarrow 2NO_2 \tag{R2.12}$$

and then dissolve in water to produce nitric and nitrous acids.

$$2NO_2 + H_2O \rightarrow HNO_2 + HNO_3 \tag{R2.13}$$

However, as we will see later, reactions such as these that are familiar at high concentrations in the laboratory do not necessarily dominate in the atmosphere, where the processes are often more subtle.

Unlike this high-temperature reaction caused by lightning, most reactions in the atmosphere occur at ambient temperature or, if they require a high energy of activation, are photochemically initiated. Such reactions form an important part of the whole subject of atmospheric chemistry. At this point we will merely glance at the reactions of a few important trace gases, paying most attention to the products of the reactions. A more detailed analysis of chemical reactions in the atmosphere will be found in Chapter 3. In the reactions dealt with above, the oxidation of lightning-produced nitric oxide to nitrogen dioxide and nitric acid illustrates the way in which chemical reactions can act as sources of gases within the atmosphere. Major reaction pathways in the atmosphere involve the oxidation of reduced gases through to more oxidised forms. This is the result of the highly oxidising nature of the atmosphere of the Earth. Thus reduced nitrogen compounds such as ammonia or nitrous oxide (NH_3, N_2O) can be oxidised, through complicated pathways to produce ultimately nitric acid or nitrate:

$$NH_3, N_2O \rightarrow NO \rightarrow NO_2 \rightarrow HNO_3, NO_3^- \tag{R2.14}$$

Sulphur compounds such as $(CH_3)_2S$, COS, CS_2 and H_2S will be oxidised in the atmosphere. The ultimate product of such oxidation is sulphuric acid or the sulphate ion. However, much of this oxidation will proceed through intermediates, of which sulphur dioxide is an important and stable one, but prone to further oxidation, i.e.

$$(CH_3)_2S, COS, CS_2, H_2S \rightarrow SO_2 \rightarrow H_2SO_4, SO_4^{2-} \tag{R2.15}$$

In the case of dimethylsulphide there are a range of stable oxidised products, such as dimethylsulphoxide, dimethylsulphone and methanesulphonic acid:

$$(CH_3)_2S \rightarrow (CH_3)_2SO, (CH_3)_2SO_2, CH_3SO_3H \qquad (R2.16)$$

Note that these more oxidised compounds are either fairly involatile (i.e. as sulphate and nitrate ions) or have a high affinity for water (i.e. sulphuric and nitric acid), so they are effectively removed from the atmosphere through rainfall or sedimentation and deposition onto the surface of the Earth. Thus, oxidation acts as a way of increasing the efficiency of removal of sulphur and nitrogen compounds from the atmosphere. Gases do not have to be oxidised right through to the most oxidised state to dissolve. Particularly noticeable is ammonia, which is already very soluble so that much of it is removed in rainfall in the lowest oxidation state, usually as the ammonium ion. To a lesser extent the moderate solubility of sulphur dioxide allows some of it to be removed in rainfall.

The oxidation of methane, the most abundant hydrocarbon in the atmosphere, proceeds via formaldehyde and offers a very significant natural source for both this gas and carbon monoxide (about 250–300 $Tg(C) a^{-1}$):

$$CH_4 \rightarrow HCHO \rightarrow CO \rightarrow CO_2 \qquad (R2.17)$$

Plant derived isoprenes and terpenes are also sources of large amounts of carbon monoxide, perhaps some 400 $Tg(C) a^{-1}$. Light is initially involved in the details of the reaction sequences above, but its role can be seen directly in the production of hydrogen and carbon monoxide through the photolysis of formaldehyde:

$$HCHO + hv \rightarrow H_2 + CO \qquad (R2.18)$$

These points will be taken up again in more detail in Chapter 3.

2.4   Sinks

If gases are to remain at a roughly constant concentration in the atmosphere then it is necessary that removal processes, representing a sink, balance production processes. As seen in the above section, chemical reactions in the atmosphere are not only sources of trace gases, but also represent sinks for the reactants. However, one might argue that these chemical processes do not represent sinks for the elements, but are simply transformations. At some point there must be a sink that removes the element from the atmosphere.

Hydrogen and helium both have an unusual sink. The molecules are light enough to achieve escape velocity ($11.3 \, \mathrm{km \, s^{-1}}$) at the high temperatures found in the exosphere that is the uppermost part of the Earth's atmosphere. A knowledge of the thermal structure of the exosphere and the related molecular velocities suggests that thermal escape would give hydrogen and helium residence times of $10^{3 \pm 0.5}$ a and $10^{8 \pm 1}$ a respectively. The observed residence time of hydrogen in the atmosphere is much shorter than this (about 10 a) because it is removed from the atmosphere through chemical reactions and in the soil by the activity of micro-organisms.

Some unreactive gases that are too heavy to escape from the Earth's atmosphere do not appear to have any sinks, i.e. the argon produced from the radioactive decay of potassium-40 (half-life 1.25×10^9 a) in the Earth has no sinks and has been accumulating since the formation of the Earth. Krypton-85, which is produced in nuclear reactors is also chemically inert, but unlike argon and helium is radioactive, and so decays with a half-life of 10.6 a to rubidium-85.

The two most important processes for the ultimate removal of atmospheric trace constituents are dry deposition and wet removal. Dry deposition is the direct transfer of gases and particles to the ground. Dry deposition is often onto wet surfaces, such as the oceans, but the word 'dry' distinguishes this mode of removal from the wet removal of atmospheric trace constituents that have dissolved in rainfall.

Soluble gases, such as sulphur dioxide, tend to dry deposit relatively quickly. This suggests that water is important in the process of deposition. Indeed it has been shown that sulphur dioxide is deposited more rapidly on dew-covered vegetation than on dry crops. It should also be remembered that the presence of crops considerably enhances the effective area available for deposition. The flux of a gas to the ground is often expressed in terms of the equation

$$F = V_{\mathrm{g}} c \tag{2.1}$$

where F is the flux, V_{g} the deposition velocity and c the concentration of the gas (usually measured at a standard height of 1 m). This equation can be used to describe the transfer of both particles and gases to surfaces. Table 2.5 gives some typical values of the deposition velocity of gases to environmental surfaces. As can be seen, even for the deposition of very soluble gases such as sulphur dioxide onto wet surfaces, the velocities are often less than a $1\,cm\,s^{-1}$. We can imagine this velocity as the rate at which a column of gas would be deposited onto the ground. This would mean that gas trapped within the first 1000 m of the atmosphere might be absorbed by the surface of the Earth in about $10^5\,s$ or about 28 h if V_g is $1\,cm\,s^{-1}$. Thus, seemingly slow deposition velocities can represent significant rates of removal. In this example the rate is equivalent to a residence time of about a day. However, such a simple model is hardly very useful because gases do not remain in such sharply defined columns in the atmosphere.

The reciprocal of V_g is known as the resistance r. If there are several different transport processes then their resistances may be summed to obtain the total resistance. For instance, in the uptake of sulphur dioxide by plants, the total resistance is given by the sum of the resistance to transfer of sulphur dioxide through the air and the boundary layer, and the surface resistance of the plant that includes the resistance offered by the cuticle of the leaf and the resistance to transfer through the stomata. If the leaf surface is wet there will be little surface resistance.

The oceans are a particularly important sink for many atmospheric gases. As seen in Table 2.5, the deposition velocity of sulphur dioxide onto the oceans can be quite high. Analysis of the concentrations of gases in seawater makes it possible to work out how undersaturated the water is with respect to these gases and to calculate the flux rates to the oceans. This is shown in Table 2.6.

Table 2.5. *Deposition velocities of a number of gases onto environmental surfaces*

Gas	Deposition velocity $(mm\,s^{-1})$	Surface
Carbonyl sulphide	0.85	Soils
	0.82	Grass – day
	0.3	Grass – night
Carbon monoxide	0.2–0.7	Soils
Hydrogen sulphide	0.15–2.8	Earth's surface
Iodine	3–22	Earth's surface
Nitrogen dioxide	0.5–6	Vegetation
Ozone	5	Vegetation
Nitric acid	20–30	Grass
Sulphuric acid	1	Grass
Sulphur dioxide	3–12	Grass
	7–10	Oceans
	8	Soils and vegetation

Deposition onto solid surfaces is also effective. Soils, or more correctly soil microbes, are able to utilise hydrogen and carbon monoxide. Thus, they can act as a sink within soils for these gases. Weathering of inorganic materials is also an important sink over long time scales (millions of years). The best known one is the removal of oxygen from the atmosphere through the oxidation of sulphide and ferrous iron. The reactions of the acid gases such as carbon dioxide and sulphur dioxide to give carbonate and sulphate minerals that represent sinks for these gases.

2.5   Global cycles and residence times

The amounts of trace materials in the atmosphere and the magnitude of their fluxes to the other reservoirs on the Earth can be summarised in terms of global cycles. Here we will be concerned only with the interchange between the atmosphere and those reservoirs that have relatively rapid rates of exchange. This usually means the Earth's surface (often the vegetated part) and the oceans. Space allows us to consider the geochemical cycling of only the most important atmospheric elements: carbon, sulphur and nitrogen.

The carbon cycle is shown in Fig. 2.1. A large part of the diagram is concerned with the pathways of reduced carbon and only the right hand side is concerned with the carbon dioxide that dominates many carbon cycles. The traditional emphasis on carbon dioxide is justified through its importance in the biosphere and the very large amounts cycled through the atmosphere each year. However, as pointed out earlier in this chapter, the importance of trace gases in the atmosphere is dependent not only on their concentration, but also on reactivity.

Formaldehyde is the most reactive gas formulated in Fig. 2.1. It is also removed in precipitation, and so has the shortest residence time of the carbon gases marked. Small amounts of long-lived gases are transported across the tropopause. Once in the stratosphere, methane, in particular, is

Table 2.6. *Fluxes of some gases to the oceans. Global flux is taken in terms of the weight of the species*

Gas	Global flux ($Tg\,a^{-1}$)
Ozone	600
Formaldehyde	10
Sulphur dioxide	8
Carbon tetrachloride	0.01
Trichlorofluoromethane	0.005

likely to be oxidised in photochemically initiated reactions. It becomes an important source of water in the dry upper levels of the atmosphere. The large amounts of methane, in the troposphere, play an important role in the carbon chemistry of the atmosphere, especially the generation of formaldehyde and carbon monoxide. The oxidation reactions of isoprenes and terpenes are another source of carbon monoxide. These natural sources of carbon monoxide are greater than the direct source from human activities.

The sulphur cycle is shown in Fig. 2.2 with the dominant biological sources of reduced sulphur. Much of this is as organic sulphides or hydrogen sulphide. Sulphur from volcanoes is usually present in more highly oxidised forms than biological sulphur, with sulphur dioxide being a common component of volcanic emissions. Most of the reduced sulphur is oxidised to sulphate in the atmosphere, although there is a significant amount of dry deposition of sulphur-containing gases, especially sulphur dioxide. The oceans represent an enormous source of oxidised sulphur.

Fig. 2.1. The global cycle of carbon. Many carbon cycles concentrate only on processes involving carbon dioxide and the biota. This part of the cycle appears on the extreme right-hand side of the figure. Assimilation and respiration will involve both terrestrial and marine processes and cycle large amounts of carbon each year. Apart from biological processes, weathering of minerals will use a small amount of the annual flux of carbon dioxide and provide a very small amount of carbon dioxide through oxidation of exposed sedimentary organic carbon. The largest sources of reduced carbon are methane from wetlands and hydrocarbons from forests. Atmospheric oxidation converts much of this to carbon monoxide, athough some of the hydrocarbons will be oxidised to other organic compounds, of which acids are the best known. Numbers are fluxes in $Tg(C)a^{-1}$.

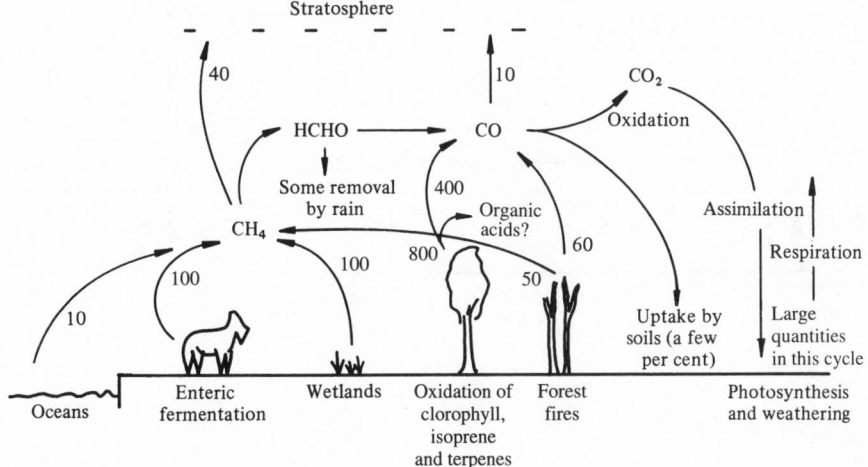

Fig. 2.2. The global cycle of sulphur. The importance of biological sources of reduced sulphur is clear in the diagram. There is also a very large source of sulphate aerosol from the sea, but much of this is returned very rapidly. Numbers are fluxes in $Tg(S)a^{-1}$.

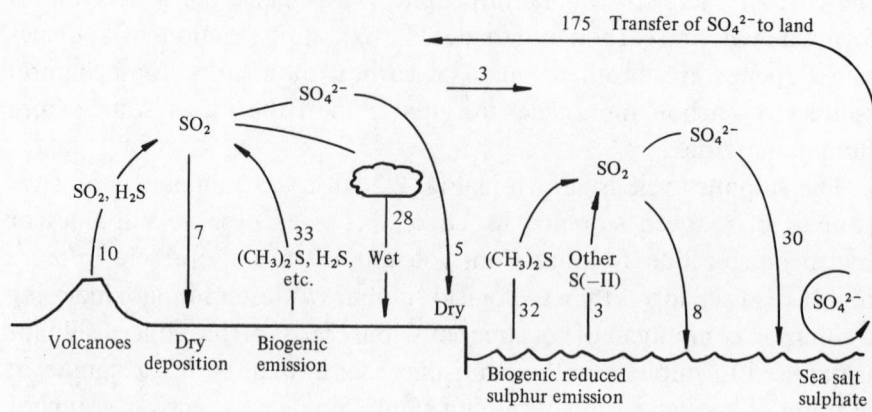

Fig 2.3. The global cycle of nitrogen. The emissions of ammonia, nitrous oxide and the nitrogen oxides are shown. About 75% of the ammonia is removed in rainfall (mostly over land) and much of the rest by dry deposition to land. About $1\,Tg\,a^{-1}$ is oxidised in the atmosphere. The nitrous oxide cycle remains unbalanced, but an important fraction goes to the stratosphere. About 30% of the nitrogen oxides are removed by dry deposition, but most are found as nitrates in rainfall. Numbers are fluxes in $Tg(N)a^{-1}$.

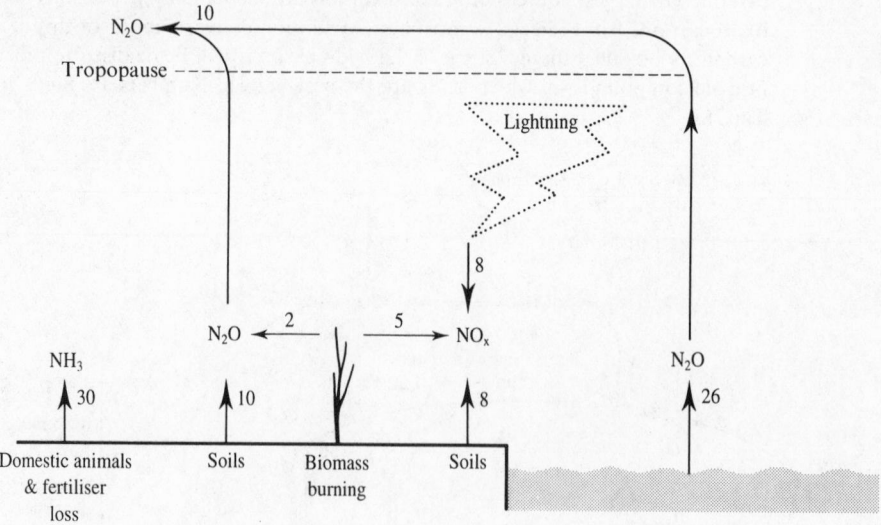

This appears as the sulphate within sea-salt. However a major part of this is recycled directly back into the sea. Even so, much is transferred to land and represents the largest of all the sulphur fluxes in the diagram. It is important to remember that this sea-salt sulphate will be near neutral in terms of acidity. It is the sulphate produced through oxidation of sulphur dioxide that generates hydrogen ions and thus represents a source of acidity in the environment.

The nitrogen cycle shown in Fig. 2.3 is not easy to balance. Each year large amounts of nitrogen and nitrous oxide are produced by denitrification. Some nitrous oxide crosses the tropopause and is oxidised in the stratosphere and this is an important sink for this unreactive gas. The transfer of gases into the stratosphere is rather slow, so nitrous oxide has a long lifetime. Nitrogen gas is fixed into terrestrial and marine biota (as proteins) and in lightning strokes (as NO_x). The biological fixation of nitrogen may be as much as $150\,Tg\,a^{-1}$. This is ultimately returned to the

Table 2.7. *Naturally occurring trace gases in the lower atmosphere*

	Residence time	Concentration (ppb)
Carbon dioxide	4 a	360 000[1]
Carbon monoxide	0.1 a	100
Methane	8 a	1 600[1]
Formaldehyde	1 d	1–0.1
Formic acid	5 d	2–0.1
Nitrous oxide	85 a	310[1]
Nitric oxide	1 d	0.1
Nitrogen dioxide	1 d	0.3
Ammonia	5 d	1
Sulphur dioxide	1–4 d	0.01–0.1
Hydrogen sulphide	24 h	0.05
Carbon disulphide	40 d	0.02
Carbonyl sulphide	2 a	0.5
Dimethyl sulphide	0.5 d	0.005
Hydrogen	2 a	550
Hydrogen peroxide	1 d	0.1–10
Methyl chloride	1.8 a	0.7
Carbonyl chloride ($COCl_2$)	2–7 d	0.05
Hydrogen chloride	4 d	0.001
Methyl bromide	1.5 a	0.005–0.025
Methyl iodide	2–5 d	0.001–0.003
Boric acid (gaseous)	5–15 d	0.05

[1]Estimated for the end of the 20th century.

atmosphere as nitrogen and nitrogen-containing gases. Ammonia is lost to the atmosphere through decomposition of proteins. A small amount is oxidised in the troposphere, but as ammonia is soluble, much of it is removed in solution or as salts such as ammonium sulphate.

Some concentrations of trace gases in the atmosphere are summarised in Table 2.7 together with their residence times. A general observation about the residence time of materials in the atmosphere has been made by Junge. He observed that the residence time of a gas is related to its variability. It is too complicated to derive this relationship in a formal way here, but it is easy to visualise. The sources of trace gases in the environment are generally less well distributed than the sinks that are often spread throughout the atmosphere as a whole. The flux of trace gases into the atmosphere is usually not related to their concentration there. On the other hand the rate of removal is very often proportional to the atmospheric concentration of the gas. These considerations mean that, if a gas is rapidly removed from the air, then it will be found at its highest concentrations near to the source and at very low concentrations elsewhere, i.e. it will show high variability. On the other hand, if a gas has

Fig 2.4. The relationship between the spatial variability of atmospheric gases and their residence time. This was originally suggested by Junge. The diagram here is modified from one by Jaenicke (1982).

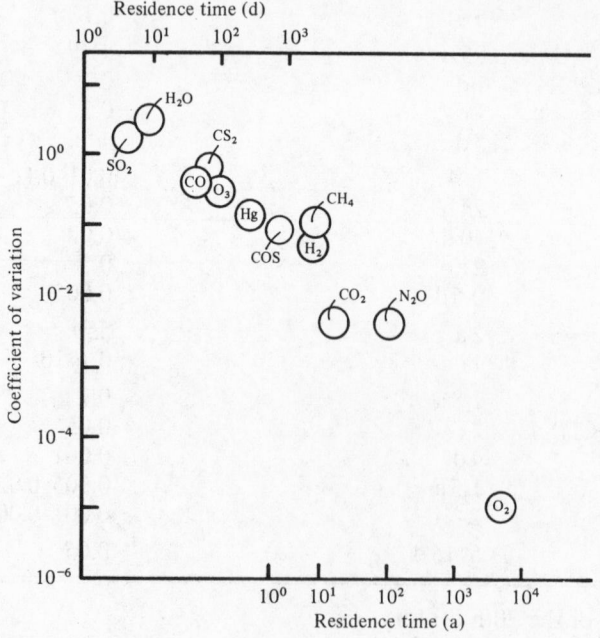

a long residence time, then it will be present throughout the atmosphere and concentration perturbations become small, i.e. the gas will have a low variability. In Fig. 2.4 the variability of various gases as a function of their residence time is shown. Sometimes this is useful for predicting the residence time of gases in the absence of any other data.

2.6   Further reading

Brimblecombe, P., Hammer, C., Rodhe, H., Ryaboshapko, A. and Boutron, C.F. (1989) Human influence on the sulphur cycle, in *Evolution of the Global Biogeochemical Sulphur Cycle*, ed. P. Brimblecombe and A. Yu. Lein, pp. 77–121. Wiley, Chichester.

Cadle, R.D. (1980) A comparison of volcanic with other fluxes of atmospheric trace constituents, *Reviews in Geophysics and Space Physics*, **18**, 746–52.

Kinzig, A.P. and Socolow, R.H. (1994) Human impacts on the nitrogen cycle, *Physics Today*, **47**(11), 24–31.

Lovelock, J.E. (1979) *Gaia. A New Look at Life on Earth*, Oxford University Press, Oxford.

3

○ ○

Gas phase chemistry of the atmosphere

3.1   Reaction rates

We have seen from the previous chapter that the atmosphere is not in a state of chemical equilibrium. The apparent stability of the atmosphere arises because it is in steady state. This situation is due to the rough balance between the input and loss to the atmosphere of many substances. As many sources and sinks of gases in the atmosphere are the result of chemical reactions, an understanding of gas phase reaction rates is important. The subject of chemical kinetics is concerned not only with the rate at which chemical processes take place, but often requires us to pay considerable attention to the mechanisms through which chemical change occurs.

Consider the oxidation of nitric oxide. This trace gas might be formed, for instance, at the shock front of a lightning bolt. Nitric oxide is oxidised by molecular oxygen according to the reaction

$$2NO + O_2 \rightarrow 2NO_2 \tag{R3.1}$$

The rate, R, at which this process occurs may be written as

$$R = - d[NO]/dt = - d[O_2]/2dt = d[NO_2]/dt \tag{3.1}$$

Here the square brackets refer to the concentration of the species contained within. In chemical kinetics, in the atmospheric context, it is usual to express concentrations as the number of molecules of gas per cubic centimetre. Note the negative signs in front of the first two differential expressions. These denote loss of NO and O_2. It is also important to realise that O_2 is lost at only half the rate of the NO loss because of the stoichiometry of the reaction. Experiments show that rate is very often a function of the concentration of the reactants. In the oxidation of nitric oxide the rate may be expressed in the form

$$R = k[O_2]^1[NO]^2 \tag{3.2}$$

k is known as the reaction rate constant. Note that the concentration terms are raised to an integer power. These exponents are known as orders of reaction. In the example above, the reaction is said to be second order with respect to nitric oxide and first order with respect to oxygen and third order overall (i.e. the sum of the individual orders). It will be noticed that the number of molecules involved in the reaction is the same as the order (i.e. second order in nitric oxide and two molecules of NO are involved). However, this is not always the case, nor does reaction order always assume integer values. As rate has the units concentration per unit time (i.e. molecule $cm^{-3}s^{-1}$) the rate constant for the oxidation of nitric oxide will have the units cm^6 molecule$^{-2}s^{-1}$.

Let us now go through reactions of various order, looking at them in a little more detail. For a fuller treatment the reader ought to refer to a standard text on chemical kinetics (see Further Reading at the end of the chapter). A first-order reaction is of the type

$$A \rightarrow products \tag{R3.2}$$

An example of this would be the decomposition of the atmospheric OCH_2CH_2OH radical:

$$OCH_2CH_2OH \rightarrow CH_2O + CH_2OH \tag{R3.3}$$

The rate constant for a first order reaction will have the units s^{-1}

$$R = -d[A]/dt = k[A]_t, (e.g. = k[OCH_2CH_2OH]_t) \tag{3.3}$$

where $[A]_t$ is the concentration of A at time t. This can be rearranged to give

$$d[A]/[A]_t = -k\,dt \tag{3.4}$$

and integrated to give

$$\ln[A]_t = -kt + c \tag{3.5}$$

The constant of integration, c, turns out to be the concentration of A at time $t = 0$, which we will symbolise $[A]_0$, and so

$$\ln([A]_t/[A]_0) = -kt \quad \text{or} \quad [A]_t = [A]_0\exp(-kt) \tag{3.6}$$

This means that the concentration of substance A will decline exponentially with time. Similar expressions relating the concentration of reacting substance as a function of time can be written for reactions of other orders. See Table 3.1.

Now let us examine a reaction that takes place in the atmosphere, i.e. the reaction of ozone with nitric oxide:

$$NO + O_3 \rightarrow NO_2 + O_2 \tag{R3.4}$$

This is a second-order reaction for which the rate can be written as

$$- d[NO]/dt = k''[NO][O_3] \tag{3.7}$$

where k'' is the second order rate constant. For this reaction $k'' = 1.8 \times 10^{-14} \, cm^3 \, s^{-1}$ at 300 K. Let us imagine the NO to be at a concentration of 0.5 ppb and ozone at concentrations now typical for the rural lower atmosphere, at about 30 ppb. Obviously these concentration units are incompatible with those of the rate constants. The conversion is simple though, because at atmospheric pressure a cubic centimetre of gas contains 2.6×10^{19} molecules (Loschmidt's number). We can obtain the concentration in the appropriate units by multiplying this number by the partial pressure of the gas. Thus, the NO and O_3 concentrations are about 13×10^9 and 7.8×10^{11} molecule cm^{-3} respectively. The rate of destruction of NO will be the product of these two concentrations and the rate constant, i.e. 1.8×10^8 molecule $cm^{-3} \, s^{-1}$. This rate is high compared with the concentration of NO, so its concentration would decline quite rapidly in a closed system. On the other hand, the ozone concentration is very much greater than the NO concentration so it will remain relatively constant. The ozone concentration may be incorporated as a constant within the rate constant and the rate expression could be rewritten as

$$- d[NO]/dt = k'[NO] \tag{3.8}$$

where k' is a first order constant given by $k''[O_3]$. This psuedo-first-order rate constant would have the value $0.014 \, s^{-1}$ at the ozone concentration under consideration. Whenever the concentration of one reactant in a second order reaction is significantly in excess of the other, the reaction

Table 3.1. *Some simple rate equations and their integrals*

Order	Differential form	Concentration relationship	Units of k
0	$- d[A]/dt = k$	$[A]_t = [A]_o - kt$	$cm^{-3} \, s^{-1}$
0.5	$- d[A]/dt = k[A]^{0.5}$	$[A]_t = ([A]_o^{0.5} - kt/2)^2$	$cm^{-3/2} \, s^{-1}$
1	$- d[A]/dt = k[A]$	$[A]_t = [A]_o \exp(-kt)$	s^{-1}
2	$- d[A]/dt = k[A]^2$	$[A]_t = [A]_o/(kt[A]_o + 1)$	$cm^3 \, s^{-1}$

can be treated as a first order process with respect to the reactant at low concentration. During the oxidation of nitric oxide, NO concentrations are expected to decline exponentially with time, as shown in Fig. 3.1. There are other ways in which the reaction order of a system may show an apparent reduction. It only requires that the concentration of one reactant remain constant. This might be because the reactant was a catalyst or because it was continually being replaced in the system.

It is convenient when reactions can be reduced to first-order systems. In such systems the concentration of the reactant will halve over a constant period of time regardless of its concentration. This half-life $(t_{1/2})$ is related to the first-order rate constant by the expression

$$t_{1/2} = \ln(2)/k' = 0.693/k' \tag{3.9}$$

The psuedo-first-order rate constant for the oxidation of nitric oxide was $0.014\,s^{-1}$, which means that the gas has a half-life of $50\,s$ in the system described above. In second-order processes the half-life is not independent of the concentration of the second-order reactant. The relationship between the second-order rate constant and half-life is given by the expression

$$t_{1/2} = 1/(k''a_0) \tag{3.10}$$

Fig. 3.1. The oxidation of nitric oxide (initially at 13×10^9 molecules cm^{-3}) by ozone at a concentration of 30 ppb (7.8×10^{11} molecules cm^{-3}).

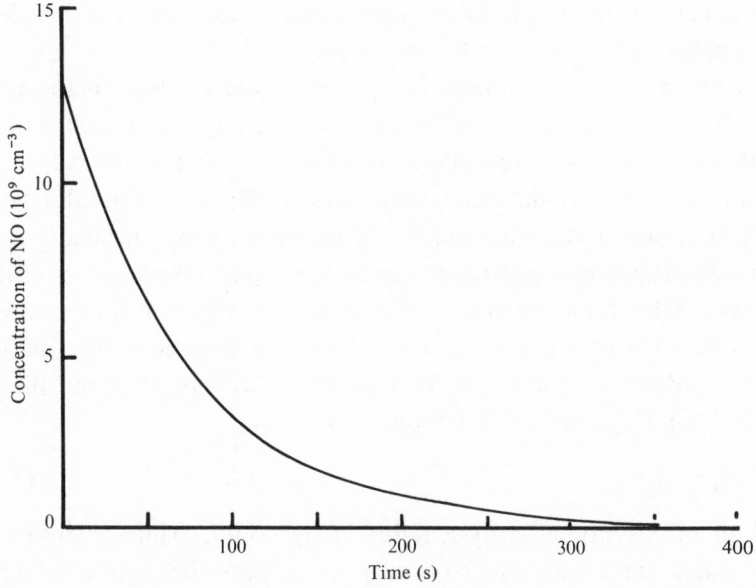

where a_0 is the initial concentration of the second-order reactant. This has important consequences that can be seen if we re-examine the oxidation of NO by O_2 (R3.1). As anyone who has ever made NO in the laboratory will know, it is converted to the brown NO_2 almost instantaneously on contact with air. The rate equation

$$- d[NO]/dt = k_3[O_2][NO]^2 \qquad (3.11)$$

where the subscript 3 has been added to remind us that the rate constant is a third-order one, can be reduced to

$$- d[NO]/dt = k''[NO]^2 \qquad (3.12)$$

because oxygen is in excess. The value of the psuedo-second-order constant is the product of the third-order rate constant, k_3 ($1.93 \times 10^{-38} \, cm^6 \, molecule^{-2} \, s^{-1}$), and the oxygen concentration (i.e. $5.46 \times 10^{18} \, cm^{-3}$, so $k'' = 1.05 \times 10^{-19} \, cm^3 \, molecule^{-1} \, s^{-1}$). Let us say that when NO is prepared in the laboratory, it has a concentration of about 1%, i.e. $2 \times 10^{17} \, molecule \, cm^{-3}$. This means that the half life calculated from equation (3.10) will be 33 s. However if we were to calculate the half-life of the same gas at concentrations typical of ambient air (i.e. 0.5 ppb or $13 \times 10^9 \, cm^{-3}$), then we would obtain a much longer half-life of 5.1×10^8 s or about 16 years. If the order of a reaction is greater than unity, then higher concentrations will give shorter half-lives. In the atmosphere, nitric oxide does not survive this long. As we have seen earlier, it reacts rapidly with other trace gases such as ozone. This is why the experience from simple laboratory experiments conducted at high concentration can sometimes be misleading.

So far this discussion of chemical kinetics would not lead us to expect great stability in the atmosphere. We have predicted a rapid decline in the concentration of nitric oxide due to reaction with ozone. However, we have totally neglected that fact that nitric oxide is continually being produced or added to the atmosphere. If the atmosphere is in steady state, then production and destruction rates are equal. We saw in earlier chapters that the residence time, or mean lifetime, was a useful parameter for describing systems in steady state. Residence time is easily obtained from first-order rate constants. In Chapter 1 we saw that the flux of material from a system was (i.e. equation (1.1))

$$F_o = A/\tau \qquad (3.13)$$

Now think of this, not as a physical flux of material (F_o) into or out of the system, but as material lost from a given volume through a chemical reaction. If the process were a first-order one and we write amount as concentration:

$$F_o = - \, d[A]/dt \tag{3.14}$$

then

$$[A]/\tau = k[A] \tag{3.15}$$

and therefore we would have

$$\tau = 1/k' \tag{3.16}$$

The assumption here is that the concentration [A] remains constant due to a continual input of material. This is true of a system in steady state, so residence times are useful for describing steady state situations. The previously calculated value for the psuedo-first-order rate constant for the oxidation of nitric oxide by ozone ($0.014\,s^{-1}$) shows that NO has a residence time of about 75 s. As we will see in Section 3.3, quite complicated systems can be understood more easily if we assume them to be in steady state.

3.2 Photochemistry

Thermal reactions of the type discussed in the last section are very familiar to students of elementary chemistry, but many key reactions in the atmosphere are photochemical. These are initiated by absorption of a photon of light rather than the intermolecular collisions that drive thermal reactions. It is possible to write these reactions just as if they were normal chemical reactions, but substituting the photon (hv) as one of the reactants, i.e.

$$NO_2 + hv \rightarrow NO + O \tag{R3.5}$$

The rate constant for such a reaction may be of second order, so the following equation can be written:

$$- \, d[NO_2]/dt = k''[hv][NO_2] \tag{3.17}$$

However, this expression is not very useful, because this second-order rate constant would vary dramatically with the energy of the photon involved in the reaction. Furthermore, as we saw in the last section, it often simplifies our understanding if we can treat reactions as first order processes. In a photochemical reaction this can be done by taking a constant flux of photons with a fixed distribution with respect to wavelength and incorporating it into a psuedo-first-order rate constant.

The rate expression then becomes

$$- d[NO_2]/dt = J[NO_2] \tag{3.18}$$

where J is the special first-order constant that embraces the absorption coefficient of the reactant, the quantum efficiency of the reaction in question and the solar spectrum and intensity at the altitude and latitude under consideration. With a little information on the spectral characteristics of the way in which the molecule absorbs light and the amount of incoming radiation, it is relatively easy to make estimates of J for atmospheric trace gases. Typical mid-latitude midday values of J_{NO_2}, for the photodissociation of NO_2, are about $5 \times 10^{-3}\, s^{-1}$, which suggests a residence time of 200 s with respect to this photochemistry.

The photochemical dissociation of NO_2 illustrates the importance of photochemistry, because one of the products is atomic oxygen. Many photochemical reactions in the atmosphere yield atoms or free radicals and these are much more reactive than molecular species. The oxygen atoms generated in the photodissociation of nitrogen dioxide can subsequently lead to the formation of ozone:

$$O + O_2 + M \rightarrow O_3 + M \tag{R3.6}$$

where M is a third body, i.e. a molecule such as N_2 that carries off excess energy that might disrupt the ozone molecule. Ozone produced by this reaction may photodissociate:

$$O_3 + h\nu \rightarrow O(^3P) \quad \text{or} \quad O_3 + h\nu \rightarrow O(^1D) + O_2 \tag{R3.7}$$

where ^3P and ^1D describe the spectral state of the atom. This reaction is significant because it is subsequently responsible for the formation of the hydroxyl radical (OH), which is probably the most important radical in the chemistry of the troposphere. If the wavelength of the photon is less than 315 nm then the oxygen atom can be produced not in the ^3P ground state, but in the excited ^1D state. While the ground state oxygen will probably recombine with an O_2 to form O_3 (i.e. no net reaction), the excited oxygen atom may be collisionally de-excited to the ground state or, more significantly, react with water:

$$O(^1D) + H_2O \rightarrow 2OH \tag{R3.8}$$

providing a source of OH radicals.

Further reactions of the hydroxyl radical can produce hydrogen atoms or the hydroperoxy radical (HO_2):

$$OH + CO \rightarrow CO_2 + H \tag{R3.9}$$

$$OH + O \rightarrow O_2 + H \tag{R3.10}$$

$$OH + O + M \rightarrow HO_2 + M \tag{R3.11}$$

$$OH + O_3 \rightarrow HO_2 + O_2 \tag{R3.12}$$

$$H + O_2 + M \rightarrow HO_2 + M \tag{R3.13}$$

Very quickly a whole suite of radicals and atoms can be generated. It is these highly reactive species that are at the heart of the gas phase chemistry of the atmosphere. The concentration of these all important radicals in the atmosphere are still only poorly known, because they are still so difficult to measure. Understandably their high reactivity means that they are found at only low concentrations. Typical daytime concentrations are about $3 \times 10^6 \, \mathrm{cm}^{-3}$ for the hydroxyl radical and $10^8 \, \mathrm{cm}^{-3}$ for the hydroperoxy radical. The high reactivity of these radicals also means that they have short residence times. The hydroxyl radical, for instance, has a residence time of about a second and the hydroperoxy radical perhaps a few minutes.

3.3 The methane cycle and continuity

We will now use the ideas of chemical kinetics and photochemistry to look at the reactions of methane in the atmosphere. Microbial activity produces atmospheric methane (R2.2), which is oxidised through intermediates such as formaldehyde. In Fig. 3.2 are shown the atmospheric reactions of methane in a schematic sense (see Table 3.2).

The notion of continuity helps in understanding the transfer of material along the various reaction pathways. It is particularly helpful when we have a complex set of reactions such as those listed in Table 3.2. Methane

Fig. 3.2. A reaction scheme for the oxidation of methane in the atmosphere. The numbers refer to reactions in Table 3.2.

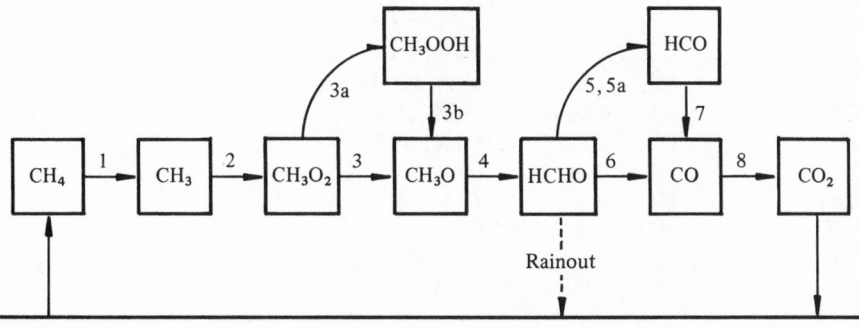

is emitted from the surface of the Earth at a mean rate of 2×10^{11} molecule cm^{-2} s^{-1}. As methane concentrations in the atmosphere remain relatively constant above each square centimetre of the Earth's surface, methane must be destroyed at this same rate, i.e. 2×10^{11} molecule cm^{-2} s^{-1}. The same could be said of the destruction and production of the methyl radical if the reactions illustrated in Fig. 3.2 were the only ones that took place in the atmosphere. Further along in the reaction scheme we see that formaldehyde is removed by four possible processes (see Fig. 3.2), this is a little more complicated, but obviously continuity requires that the sum of the fluxes through these four pathways be equal to the production rate. This is necessary for the system to remain in steady state. We can look back through the reactions in Fig. 3.2 and equate the destruction of formaldehyde to the production of methane at the surface of the Earth or its destruction in the atmosphere. So we can write

$$-d[CH_4]/dt = -d[HCHO]/dt \qquad (3.19)$$

Where rain-out is not very significant, this equation can be rewritten as

$$k_1[OH][CH_4] = k_5[HCHO][OH] + J_{5a}[HCHO] \\ + J_6[HCHO] \qquad (3.20)$$

The concentration of CH_4 is 1.72 ppm or 4.47×10^{13} molecule cm^{-3} and let us take that of the OH radicals to be 10^6 molecule cm^{-3}. The rate constant k_1 is 8×10^{-15} molecule cm^3 s^{-1} and the destruction rate of methane can be estimated as 3.58×10^5 molecule cm^{-3} s^{-1}. Furthermore, equation (3.20) can be rearranged to give

$$[HCHO] = k_1[OH][CH_4]/(k_5[OH] + J_{5a} + J_6) \qquad (3.21)$$

so that the concentration of formaldehyde can be estimated as 4.3×10^9

Table 3.2. *A reaction sequence for the oxidation of atmospheric methane*

(1)	$CH_4 + OH \rightarrow CH_3 + H_2O$
(2)	$CH_3 + O_2 + M \rightarrow CH_3O_2 + M$
(3)	$CH_3O_2 + NO \rightarrow CH_3O + NO_2$
(3a)	$CH_3O_2 + HO_2 \rightarrow CH_3OOH + O_2$
(3b)	$CH_3OOH + h\nu \rightarrow CH_3O + OH$
(4)	$CH_3O + O_2 \rightarrow HCHO + HO_2$
(5)	$HCHO + OH \rightarrow HCO + H_2O$
(5a)	$HCHO + h\nu \rightarrow HCO + H$
(6)	$HCHO + h\nu \rightarrow CO + H_2$
(7)	$HCO + O_2 \rightarrow CO + HO_2$
(8)	$CO + OH \rightarrow CO_2 + H$

molecule cm^{-3} (where $k_5 = 1.3 \times 10^{-11}$ cm^3 s^{-1} and $J_{5a} + J_6 \simeq 4.5 \times 10^{-5}$ s^{-1}). This estimate of concentration is close to the mean concentration of 6.17×10^9 molecule cm^{-3} observed in the unpolluted lower atmosphere.

Let us now apply the notion of continuity to the formation of carbon monoxide by oxidation of methane. Carbon monoxide in the atmosphere arises from many sources, but it is easy to assess the size of the methane source. Despite the complicated path of oxidation from methane, if we neglect the small loss due to rainout of formaldehyde (and perhaps to a lesser extent rain-out of methyl hydroperoxide, CH_3OOH) carbon monoxide should be formed at the same rate as methane is released into the atmosphere, i.e. 2×10^{11} cm^{-2} s^{-1} or about 700 Tg(C) a^{-1}; this is a little larger than the amount produced by human activities (450 Tg(C) a^{-1}).

Carbon monoxide itself will react with hydroxyl radicals in a process that represents an important sink for the radical:

$$OH + CO \rightarrow CO_2 + H \qquad \text{(R3.14)}$$

3.4 Atmospheric chemistry of the hydroxyl radical

The principles outlined in the previous section can be applied to the reactions of other trace components in the atmosphere. Many gases that we traditionally think of as stable react quite readily with radicals. This section concentrates mainly on reactions of trace gases with the hydroxyl radical that exert a profound effect on the composition of the atmosphere.

The reaction of the hydroxyl radical with methane showed that these radical reactions are frequently initiated by abstraction of hydrogen. We can also see this with a nitrogen-containing gas such as ammonia:

$$NH_3 + OH \rightarrow NH_2 + H_2O \qquad \text{(R3.15)}$$

with a number of subsequent steps being possible, e.g.

$$NH_2 + NO_2 \rightarrow H_2O + N_2O \qquad \text{(R3.16)}$$

This mechanism accounts for little of the ammonia removal from the Earth's atmosphere, because most ammonia dissolves in liquid water.

Nitrogen dioxide, the product of many oxidation reactions, is a very important compound in the atmosphere. The hydroxyl radical may convert this through to nitric acid that is very soluble in water and can be removed effectively by rain:

$$NO_2 + OH + M \rightarrow HNO_3 + M \qquad \text{(R3.17)}$$

However, it is important to note that the reaction is a hydroxyl radical addition and different from the abstractions we have seen so far.

Biologically produced sulphur gases are emitted as sulphides. Hydrogen sulphide can undergo hydrogen atom extraction:

$$OH + H_2S \rightarrow HS + H_2O \tag{R3.18}$$

The HS is oxidised through to SO_2 by subsequent reactions, although they are not well understood. The sulphur dioxide produced by this and other processes can also be oxidised by OH:

$$SO_2 + OH + M \rightarrow HSO_3 + M \tag{R3.19}$$

$$HSO_3 + O_2 \rightarrow HO_2 + SO_3 \tag{R3.20}$$

$$SO_3 + H_2O \rightarrow H_2SO_4 \tag{R3.21}$$

The radical reactions lead to sulphuric acid, the ultimate product of the oxidation of atmospheric sulphur compounds.

The major biological emission of sulphur as dimethylsulphide can react either by abstraction:

$$CH_3SCH_3 + OH \rightarrow CH_3SCH_2 + H_2O \tag{R3.22}$$

$$CH_3SCH_2 + O_2 \rightarrow CH_3SCH_2O_2 \tag{R3.23}$$

$$CH_3SCH_2O_2 + NO \rightarrow CH_3S + NO_2 + HCHO \tag{R3.24}$$

or by addition:

$$CH_3SCH_3 + OH \rightarrow CH_3SOH + CH_3 \tag{R3.25}$$

$$O_2 + CH_3SOH \rightarrow CH_3SO_3H \tag{R3.26}$$

The addition process is not well understood, but the product of the reaction above is methanesulphonic acid, which is a well-known organosulphur compound in the atmosphere. It is very soluble and dissolves in aqueous solutions in the atmosphere (e.g. raindrops), where it is resistant to further oxidation. This relative stability means that it is a useful marker for marine air masses that once contained relatively high concentrations of dimethylsulphide.

Hydroxyl radicals can react to form hydrogen peroxide:

$$OH + OH + M \rightarrow H_2O_2 + M \tag{R3.27}$$

but in most situations the hydroperoxide radical provides a more efficient route:

$$HO_2 + HO_2 \rightarrow H_2O_2 + O_2 \qquad\qquad (R3.28)$$

Hydrogen peroxide is very soluble in water and a strong oxidising agent, so it plays an important role in oxidation processes within water droplets in the atmosphere.

3.5   The nitrate radical

The rapid reaction of the hydroxyl radical with many trace gases in the atmosphere places it at the centre of much atmospheric chemistry. Yet there are important reasons why it does not totally dominate the chemistry of the atmosphere. Perhaps the most significant limitation to the role of the hydroxyl radical is the fact that its concentrations are so low at night.

At low light levels, the nitrate radical proves to be an important trace entity. It can form thus

$$NO_2 + O_3 \rightarrow NO_3 + O_2 \qquad\qquad (R3.29)$$

During the day concentrations are low because it photolyses effectively even at long wavelengths ($<630\,nm$):

$$NO_3 \rightarrow NO_2 + O \qquad\qquad (R3.30)$$

and high daytime concentrations of NO (from NO_2 photolysis) may also reduce its concentrations:

$$NO_3 + NO \rightarrow 2NO_2 \qquad\qquad (R3.31)$$

At night the NO_3 radical can act as a hydrogen atom abstractor in much the same way as OH. For instance an alkane such as methane could be attacked:

$$CH_4 + NO_3 \rightarrow CH_3 + HNO_3 \qquad\qquad (R3.32)$$

although it must be admitted that the reactions with alkanes are not particularly fast. Reactions with aldehydes, alkenes, terpenes and aromatic compounds are more effective.

The nitrate radical also reacts:

$$NO_3 + NO_2 + M \rightarrow N_2O_5 + M \qquad\qquad (R3.33)$$

Dinitrogen pentoxide (N_2O_5) can react heterogeneously with water to give nitric acid, or decompose to give the original reactants. One interesting possibility is that the nitrate radical can generate HO_2 at night through the reaction sequence:

$$NO_3 + HCHO \rightarrow HCO + HNO_3 \tag{R3.34}$$

$$HCO + O_2 \rightarrow CO + HO_2 \tag{R3.35}$$

It becomes easy to see why the nitrate radical has the potential for playing such an active role in the night-time chemistry of the troposphere.

3.6   Conclusions

In simple terms one can see the hydroxyl radical controlling the daytime chemistry and the nitrate radical that at night. This is generally true, but there are other reactive entities that can drive the chemistry of particular trace gases. The alkenes are effectively attacked by O_3 or the ground state oxygen atom. Formaldehyde, in particular, reacts very rapidly with the hydroperoxy radical.

This chapter has outlined some of the more important sequences of reactions in the atmosphere. The ideas developed here will be useful in understanding the way in which pollutants perturb the chemistry of the troposphere (Chapter 6) and stratosphere (Chapter 8). They will also prove helpful in unravelling the more exotic chemistry taking place on other planets (Chapter 9).

3.7   Further reading

Harrison, R. M., de Mora, S. J., Rapsomanikis, S. and Johnston, W. R. (1991) *Introductory Chemistry for the Environmental Sciences*, Cambridge University Press, Cambridge.
Laidler, K. J. (1987) *Chemical Kinetics*, Harper Row, New York.
Wayne, R. P. (1991) *Chemistry of Atmospheres*, Oxford University Press, Oxford.

4

○ ○
Aerosols

A system of small particles or liquid droplets suspended in a gaseous phase is called an aerosol. Popular usage would have it that aerosols are the sprays derived from aerosol spray cans. Many of these are aerosols, because the spray consists of fine droplets dispersed in the air. Aerosols can be thought of as the opposite to foams which are dispersed systems of gas in a liquid or solid phase. Aerosol particles have to be very small, not too much larger than a micrometre in radius, to remain suspended in the air for long periods.

In spite of the microscopic size of aerosols, their presence in the air is discernible because they give a hazy appearance to distant objects. The fact that aerosols are easily perceived means that dusts and smokes have been written about since classical times. Early writers leave us in no doubt that they regarded them as undesirable. They were considered detrimental to health and annoying because they discoloured buildings and lowered visibility. In more recent times, perhaps because of an increased appreciation of perspective, mistiness in views of distant objects was thought to lend an attractive quality to landscapes. Haze in the Blue Mountains of Australia or the Great Smoky Mountains of North America occurs frequently enough to be embodied in their names. The haze over forested areas (see also Section 2.1) has often been attributed to the oxidation products of the terpenes that evaporate from the trees on hot days. However, recent measurements in North America suggest that much of the haziness over forests in the present day is due to pollution from sulphate aerosols.

4.1   The origin of aerosols

There are two main sources of fine particles in the atmosphere: primary particulate material that is directly derived from dispersal of solids from the Earth's surface and secondary particulate material, which forms because of chemical reactions in the atmosphere.

Primary particulate material

Many viable particles are found in the air. They include pollens, spores, micro-organisms and insects (together with insect fragments). Pollen is a particularly troublesome component of the air to many people who contract hay fever in seasons when the pollen concentration is high. Grains and fragments of pollen lie in the size range 5–25 µm radius.

In recent years there has been a growth in interest in the presence of biological particles in indoor air. Outbreaks of serious diseases such as *Legionella* have been linked to problems with air-conditioning systems. These incidents have heightened awareness of the enormous range of biological material found in the air of dwelling places. This would include vegetable dusts, hair, skin, feather and dander from animals, and a range of bacteria, fungi and other micro-organisms (for further reading see Leslie and Lunau (1992)).

Outdoors biological processes also produce a range of non-viable particles. The decay of leaf litter gives rise to quite large amounts of particulate material. Plants have been observed to emit particles directly into the atmosphere. This direct emission yields needle-shaped particles (200 nm × 30 nm), which contain high concentrations of metals such as zinc, cadmium and lead. The biologically mediated volatilisation proves to be important, on a global scale, for mercury, arsenic and selenium. These emissions from plants may be in both the gas and the particulate phase, but are often associated with the production of spores, waxes and other plant particles (see Table 4.1).

A more familiar source of particulate materials from vegetation is the forest fire. Particles generated during the combustion of vegetation are quite small, with 0.05 µm being suggested as an average radius. Fires can

Table 4.1. *Global production of vapour phase and particulate trace elements by plants (from Nriagu (1989))*

	Volatiles ($Gg\,a^{-1}$)	Particles ($Gg\,a^{-1}$)
Arsenic	0.26	1.3
Cadmium	0.15	0.04
Cobalt	0.52	0.06
Copper	2.6	0.32
Mercury	0.02	0.61
Manganese	27	1.3
Nickel	0.51	0.1
Lead	1.3	0.2
Antimony	0.2	0.04
Selenium	1.12	2.6
Vanadium	0.92	0.13

be strong local sources of particulate matter, yielding up to several tonnes of smoke per hectare $(10^4 \, m^2)$. On a global basis they may produce between 6 and 30 Tg smoke per year as elemental carbon.

An important source of particles is the suspension of dusts from dry areas, where concentrations of almost a milligram of solid per cubic metre of air may be found. This is an order of magnitude higher than the concentration of soot in polluted urban air. Important geographical areas for this dust production are the Sahel, the deserts of Asia and some parts of the USA. Fig. 1.4(c) showed the frequent haze of various maritime areas.

Studies of the mechanisms of dust generation have often been made because of an interest in sandstorms and sand dune formation. Much of the particulate material suspended in this process is very coarse, but the suspension process can give rise to a finer long-lived atmospheric aerosol. Dust generation begins with lateral transport of material, or saltation, in which the wind moves large particles along such that they can develop sufficient lift to become airborne. However, they are usually so large that they can only remain in suspension for a few seconds before dropping to the ground. A wind velocity of some $0.2 \, m \, s^{-1}$ is required to set the particles at the surface in motion. This represents a wind speed of a few metres per second at the normal height of measurement (1–10 m). The particles in motion under such conditions are roughly 0.1 mm in diameter. Collision with the ground sets smaller particles in motion. These collisions can also lead to fragmentation and the production of finer material. Once the finer material is in suspension it stays aloft for much longer periods. Observations suggest that high concentrations of fine particulate matter in the air above deserts are not limited to the periods of intense storms.

The material generated in this way should resemble the surface from which it was derived in terms of mineralogy. In Fig. 4.1 the excellent correlation between the composition of airborne dusts in the Sudan and the local crustal composition is shown. The dusts can be transported long distances and a characteristic mineralogy will often allow subsequent identification. The yellow haze found in the Far East that sometimes colours snow packs of the Arctic usually arises from continental material driven aloft in winds over the Asian deserts. Reddish deposits of dust from the Sahara are occasionally found in Europe. A particularly spectacular fall occurred in February 1903 and more recently widespread dust deposits were observed in the UK in November 1984.

A further source of terrigenous primary particulate material is the eruption of volcanoes. As with the sources of desert dust, volcanoes are restricted in geographical location. In addition, eruptions are usually sporadic. Nevertheless, volcanoes represent a source of very large amounts of particulate material in the atmosphere. The particle size range found in volcanic material is wide. The biggest of these particles, which would not normally remain in the air for any length of time, can travel some distance because of high ejection velocities. The concentrations of aerosols can be enormous. In a brown cloud of dust from the eruption of St Augustine in Alaska during February of 1976, aerosol concentration was estimated to be $0.1 \, g \, m^{-3}$. The eruption of Mount St Helens also led

Fig. 4.1. Elemental composition of the soil compared with that of airborne dust in Sudan (data from Pinkett *et al.* (1979)).

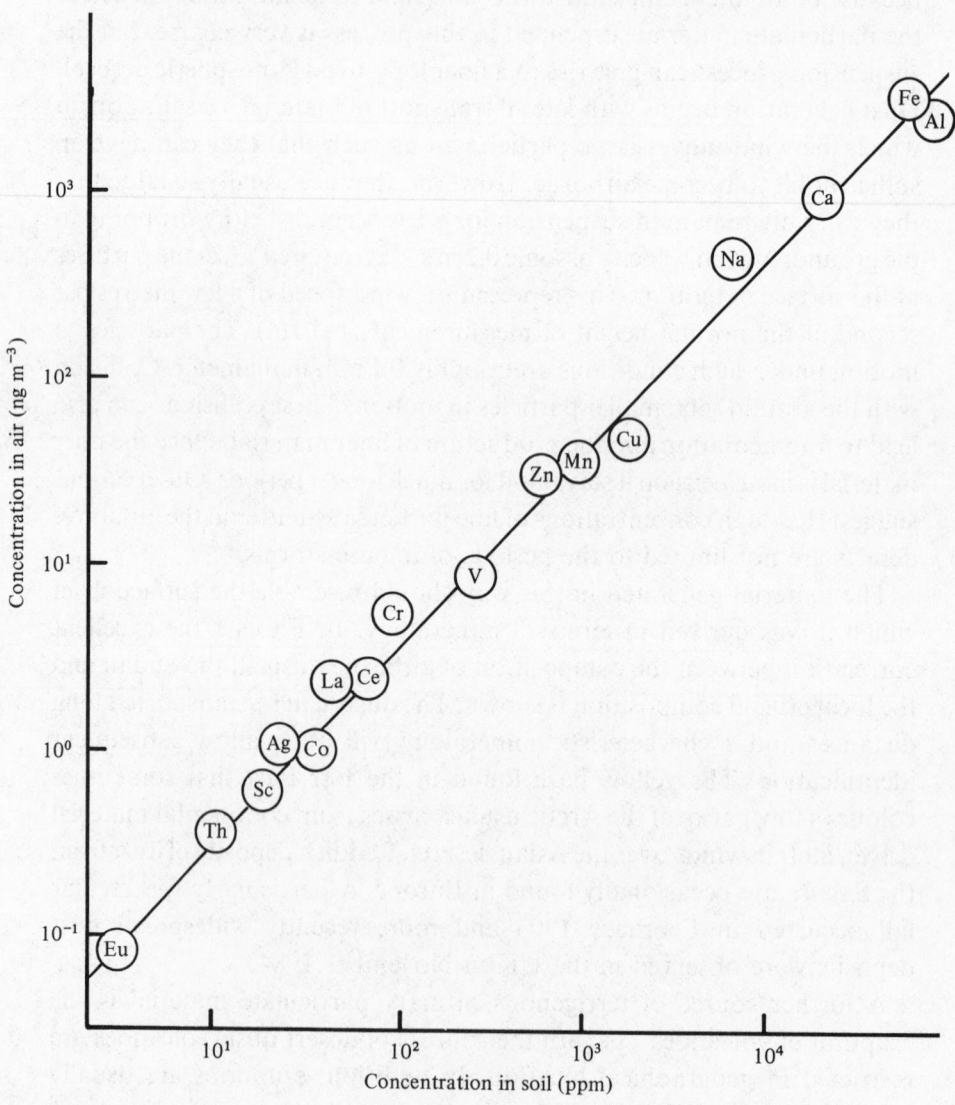

to large increases in the concentrations of atmospheric aerosol. The concentrations of larger particles, i.e. those in the 1–10 μm range, were enhanced a thousand-fold above their normal levels downwind of the volcano. However, the concentration of particles less than 1 μm in diameter remained close to ambient, which emphasises that the primary particles emitted by volcanoes are relatively coarse.

Sea surfaces are large and widely distributed sources of particulate material. It was once thought that wind-blown spray and foam were the source of the suspended material, but the size of the particles from this source is too great for them to have long residence times in the air. The generation of aerosols by the oceans was thoroughly investigated in the 1950s and since then a clearer picture of the maritime source has emerged. Bubbles from breaking waves and biological activities are now considered to the major source of the sea salt aerosol. It can be seen from high-speed cinematography of the bubble-bursting process that there are at least two types of microscopic droplets (see Fig. 4.2).

(i) Small droplets arise from disintegration of the film that forms the cap of the bubble. This process is more important in the case of larger bubbles. A bubble of radius 1 mm may give a single film-droplet while a 3 mm bubble may give more than ten.

Fig. 4.2. Production of film and jet drops from bursting bubbles.

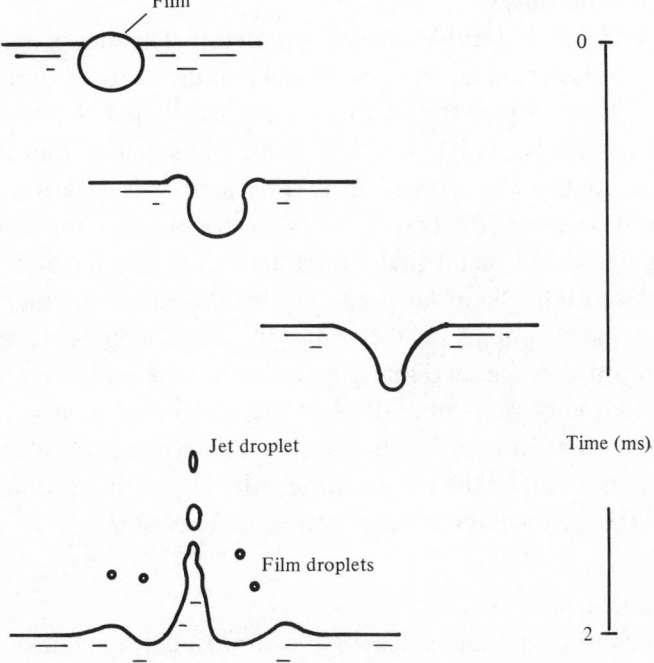

(ii) Larger droplets are derived from the central jet that forms as the bubble cavity collapses, and this is likely to be the dominant source of droplets from the smaller bubbles of radius 0.2 mm that are generated by breaking waves. Here, the smaller the bubble, the larger the number of jet droplets. About five result from a 0.5 mm bubble, yet only a single droplet is produced by a 3 mm bubble.

The production of bubbles occurs in the whitecaps, which cover about 2% of the sea surface. In these regions the flux of bubbles to the surface is about $2 \times 10^6\,\mathrm{m^{-2}\,s^{-1}}$. Naturally, the highest rates of production are to be found in windy conditions. Once the seawater droplets formed by the bursting bubbles are in the air they can evaporate leading to concentrated saline droplets or minute crystals of airborne sea salt. A bubble of radius 1 mm is likely to produce a few film droplets from the disintegrating cap. They will have radius 3–15 μm and contain 4–50 pg of salt that would give a dry particle of 0.8–4 μm. The central droplet, from the bubble of radius 1 mm, will have a radius of about 10% of the bubble so it will contain 0.15 μm of salt (producing a dry particle of 25 μm radius). The size spectrum of bubbles and droplets produced and the flux of material from the surface of the ocean are shown in Fig. 4.3. The exact magnitudes of these fluxes are still poorly known for the ocean, as most observations remain laboratory-based.

Meteoritic debris is another source of material. The flux of material of radius > 100 μm is about $10^{-9}\,\mathrm{m^{-2}\,s^{-1}}$ and that of radius > 1 μm is about $10^{-4}\,\mathrm{m^{-2}\,s^{-1}}$ at the top of the Earth's atmosphere. These particles will be depleted in materials of high volatility. This still suggests that the total contribution to the atmosphere from this source is relatively small. However, just occasionally there is a very large meteorite impact capable of injecting significant quantities of terrestrial dust into the atmosphere. The dust placed into the atmosphere, in this less direct manner, by the Tunguska meteorite impact in Siberia in 1908 caused a series of impressive sunsets. It has been suggested that gigantic meteorite impacts in the past have generated enormous quantities of dust, sufficiently large to have blocked off sunlight for many years, with as a consequence the elimination a substantial fraction of the life on the Earth. The global production of particles in the atmosphere is summarised in Table 4.2.

Secondary particulate material
Chemical reactions taking place in the atmosphere can give rise to particles. Nitrate-containing aerosols are formed by the oxidation of

Table 4.2. *Sources for particulate material in the atmosphere*

Source	Global flux (Tg a^{-1})
Primary	
Forest fires	35
Dust	300
Sea salt	1000
Volcanic dust	50
Meteoritic dust	1
Secondary	
Sulphate production	150
Nitrate production	250
Organic particles	100

Fig. 4.3. The flux of variously sized jet and film drops from the bubble spectrum shown in the lower part of the figure (from Blanchard (1983)).

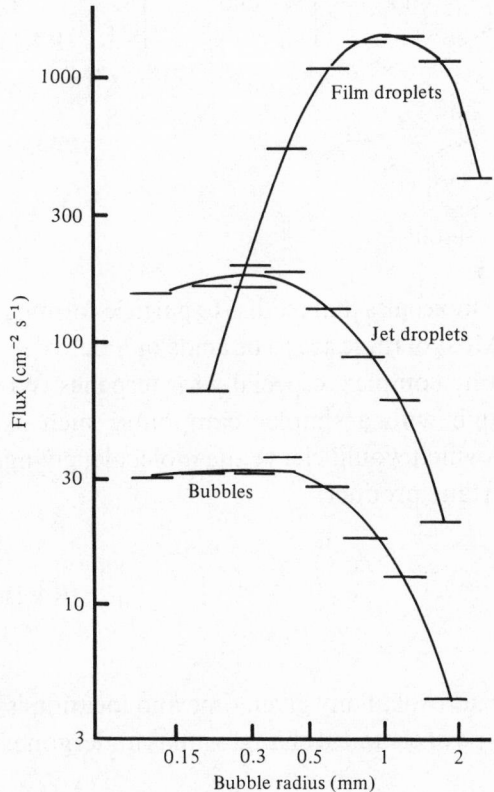

nitrogen oxides. In a similar way sulphate aerosols can form from sulphur dioxide in the atmosphere. Volcanoes are a large and important source of this sulphur dioxide. It is also possible for volcanic sulphur to be injected into the stratosphere during intense eruptions. Production of a secondary sulphate aerosol in the stratosphere has a greater impact on climate than has dust, because it has a much longer lifetime than dust. Ammonium bisulphate or sulphate aerosols are formed by the reaction of atmospheric ammonia with sulphuric acid. This sulphuric acid will have been produced by the oxidation of natural and industrial sulphur dioxide.

Biological processes are understandably the most important in producing organic precursors to secondary aerosol particles. The oxidation of a terpene, such as pinene, by photolysis in the presence of $NO_x(a)$ and ozone (b) gives a range of products (Warneck, 1988):

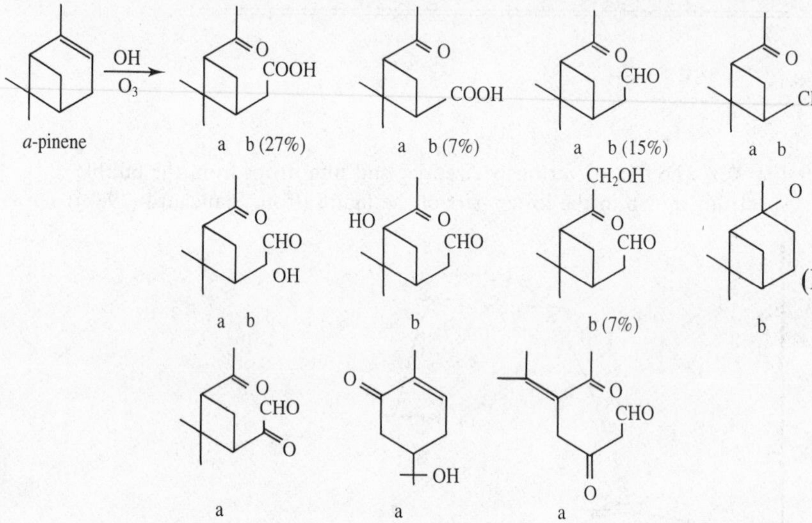

(R4.1)

It is usually the larger unsaturated molecules that oxidise to particle-forming compounds in the atmosphere. Most of these are to be acids or aldehydes. The reaction sequence can be quite complex, especially for terpenes. We can illustrate the process, though with a simpler compound such as cyclohexene, ozone attack upon which would cleave the molecule, giving 6-oxohexanoic acid as an important product:

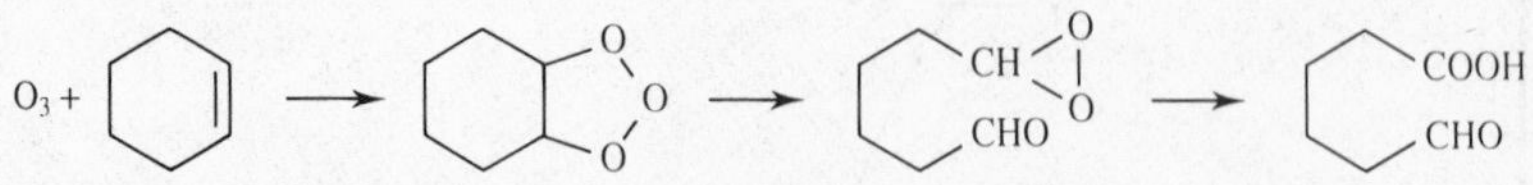

(R4.2)

Composition
The chemical composition of the aerosol at any given time and location is a product of its sources and the type of chemical changes it has undergone.

Insoluble aerosols, though not immune to change, are more resistant than soluble particles. The maritime aerosol is dominated by the presence of sea salt, but, as we shall see later, this can be modified. The continental aerosol is more likely to be influenced by dust (predominantly silicon, aluminium and iron oxides). Ammonium sulphate (or ammonium bisulphate) at concentrations that can be in excess of $10 \ \mu g \ m^{-3}$ is a major soluble component over the land, although chloride content increases markedly in coastal areas. As mentioned, ammonium sulphate has its origin in the reaction between sulphuric acid and ammonia. Even over the sea, ammonium sulphate contributes to the aerosol, although here sulphate can frequently be found as sulphuric acid.

Organic compounds are also found as atmospheric aerosols. The ether-soluble compounds detected include organic acids and bases, phenols and aliphatic and aromatic hydrocarbons. They have a total concentration, in remote areas, of about $1 \ \mu g \ m^{-3}$. These organic compounds are necessarily low-volatility ones, so we find the higher alkanes for instance. A preference for the odd-numbered alkanes indicates vegetation to be a source for this organic material. Aliphatic alcohols and carboxylic and dicarboxylic acids are also common constituents of aerosols, which have probably been derived, directly or indirectly, from plant waxes. Various terpenoids derived from the essential oils of plants are also to be found in atmospheric aerosols. Some nitrogen-containing organic compounds, such as amino acids, are also present in particulate material. They are presumably of biological origin, but the processes through which these get into the atmosphere are not yet understood.

Forest fires, a prolific source of elemental carbon, can also produce particles containing other organic and inorganic materials. The oxidised organic compounds, such as acids, are found together with a range of soluble salts of sodium and potassium. The anions to such salts would be chlorides, sulphates, nitrates and even phosphates. Although soot has an obvious product of combustion, long-lived small carbon particles have only been studied in the last two decades. The forest fire is a natural source, but today even in quite remote places, long-range transport of pollutants derived from the internal combustion engine make a major contribution. Carbon particles are effective at both scattering and absorbing light, so they can affect visibility and alter the radiative properties of the atmosphere. Their presence has been considered with regard to the pollutant hazes over large cities. In the 1980s much research was done on the effect of carbon particles injected into the atmosphere from fires caused by the use of nuclear weapons as some calculations suggested a global 'nuclear winter' could result. Similar fears were expressed when enormous oil fires burnt at the end of the Gulf War in 1990. However, the smoke rose only a few kilometres and had a short residence time, so the cooling effects were localised to within a few hundred kilometres.

4.2   Size spectra of atmospheric particulate/ aerosol material

In the section above, the size of the particulate material under discussion has sometimes been mentioned, but this is so important that descriptions without a thorough discussion of particle size are very restricted. Particles must be small just to allow them to stay airborne, but size governs a whole range of properties possessed by particulate material. Aerosols are rarely of a uniform size, so it is necessary to consider their size spectra.

Size is difficult to define for atmospheric particles, because usually we do not know their shape. Often they are treated as spheres, because the terminal velocity of small spheres, v_t, is easily derived from Stokes' Law:

$$v_t = 2r^2 \rho_p g/(9v) \tag{4.1}$$

where ρ_p is the density of the particle and v the viscosity of the fluid. Strictly speaking, the difference between the particle density and the density of the air should be used rather than particle density, but the density of air is usually small enough to neglect. Besides variation in shape, density is also likely to be variable, so sometimes it is convenient to define an equivalent aerodynamic radius. This is the radius of a sphere of unit specific gravity having the same falling velocity as the particle under question. Particles in the atmosphere follow Stokes' law closely over the size range 15–30 µm diameter. Outside this range it is necessary to apply corrections (Hesketh, 1977).

The types of particles found in the atmosphere are shown in Fig. 4.4. The size range normally considered covers many orders of magnitude so representation is easiest on a logarithmic scale. Haze particles are close to the wavelength of light in size, hence their optical activity. However, the smallest particles of interest to us are the Aitken nuclei. They range from about 4×10^{-3} µm upwards in radius and the upper limit is usually put at about 0.1 µm. Most of these are secondary particles produced from chemical reactions in the atmosphere. They are often called condensation nuclei, but this is not a good term because although they are often detected by condensation techniques, their effects in the atmosphere are mostly electrical. Normally they are not involved in cloud formation. The rate of diffusion of such small particles is large so that they coagulate according to the formula

$$- \mathrm{d}N/\mathrm{d}t = 8\pi DrN^2 = k''N^2 \tag{4.2}$$

where N is the number of particles, t the time, r the radius and D the diffusion constant. The equation applies for monodisperse particles (uniformly sized) and we should note that the rate is essentially of second order so the half-life will depend on the concentration such that $t_{1/2} = 1/(k''N)$ (see Section 3.1). However, coagulation will soon change the sizes of the particles. The diffusion constant is given by the relationship originally put forward by Einstein and modified by Cunningham:

$$D = kT(1 + AL/r)/(6\pi vr) \tag{4.3}$$

where k is the Boltzmann constant (1.38×10^{-23} J K^{-1}), T the absolute temperature, v the viscosity of the medium (18.27×10^{-6} N s m^{-2}), A the Stokes–Cunningham constant ($1.257 + 0.4 \exp(-1.1r/L)$) and L the mean free path of gas molecules (6.53×10^{-8} m). If we determine the rate of coagulation for particles of 0.3×10^{-8} m radius at a concentration of 10^9 m^{-3} (i.e. 1000 cm^{-3}) then we find that the half-life is about a day. This shows that coagulation of very small particles is moderately rapid. However, remember that coagulation leads to particles of

Fig. 4.4. Nomenclature for atmospheric particles. Much of the early work on this was done by C. E. Junge (1963). The three classes at the bottom of the diagram separate the particles on the basis of their origin. Initial nuclei are so small that they have to be formed by condensation of vapours or chemical reactions. These can coagulate to form slightly larger particles. The largest particles arise from dispersal of materials at the Earth's surface.

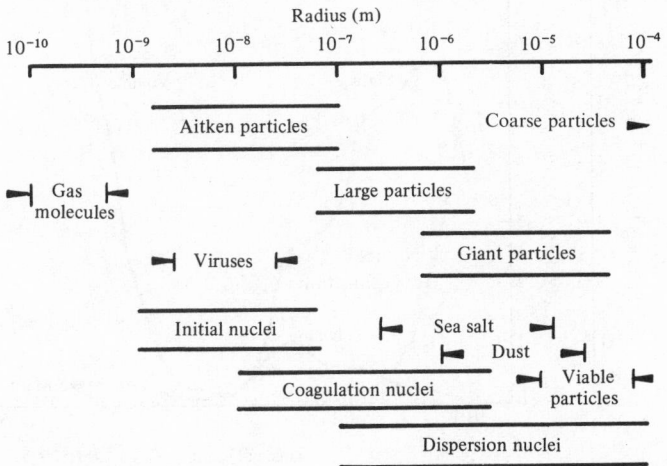

greater radius, so the process eventually comes to a standstill.

The large and giant particles, which appear in Fig. 4.4 and Fig. 4.5 between the radii 0.1 and 10 μm, make up the majority of the mass of particulate material in the atmosphere. Large particles, especially the smaller ones in this class, are probably formed by coagulation. Their diffusion is almost negligible, so they act as receptors for smaller particles. Many giant particles are silica, calcium carbonate, sodium chloride or ammonium sulphate. The last two of these are probably formed by evaporation of droplets. Regardless of their origin large particles fall out rather rapidly, with speeds of a few centimetres per second.

Measurements of particulate material in the atmosphere are normally displayed in terms of a size distribution function. This function is often seen plotted as the number of particles over a given size range that are to be found in a cubic centimetre of air. If the size range of the interval is made very small then the concentration can be written as the differential, dN/dr or $dN/d\log(r)$. Measurements have shown that $dN/d\log(r)$ may be related to the radius, r, by the equation

$$dN/d\log(r) = Cr^{-\beta} \tag{4.4}$$

A graphical representation of a particle size and various distribution functions is shown in Fig. 4.5. It is seen that, at very small particle size, the

Fig. 4.5. Concentration–size distributions for an aerosol adapted from measurements made by Willeke and Whitby in the Denver City Maintenance yard in 1971 (Willeke and Whitby, 1975).

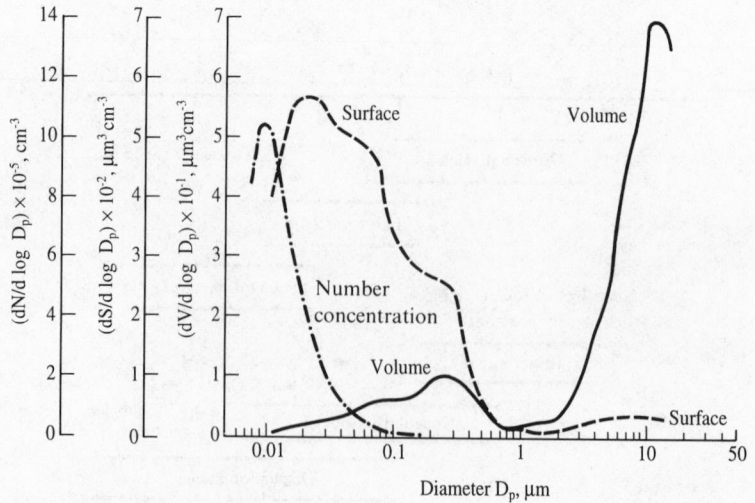

concentration of particles is relatively low. This is due to the higher rate of coagulation for smaller particles. However the number concentration peaks, so that the largest concentrations of particles are to be found in the smaller size ranges. Concentration soon begins to decline rapidly with size. The decline of concentration with changing size is often approximately linear on logarithmic graphs, i.e. β in equation (4.4) is constant. Over the larger size ranges, β is about 3. This value of β suggests that the particles were formed through fragmentation processes. In the middle size ranges it may be close to 2, indicating that the particles may have been formed through accumulation. Much of the volume, and therefore mass, is associated with the largest particles.

Slinn has suggested that the differences between aerosols with different origins are clearer if plotted as a surface area distribution rather than a number distribution. Such a distribution function is shown for particulate material from many sources in Fig. 4.6. The continental aerosol shows a bimodal distribution, the smaller particles being the initial nuclei and the larger ones derived from accumulation of these small particles. Sea salt

Fig. 4.6. The area distribution proposed by Slinn (1983) for different types of atmospheric particles.

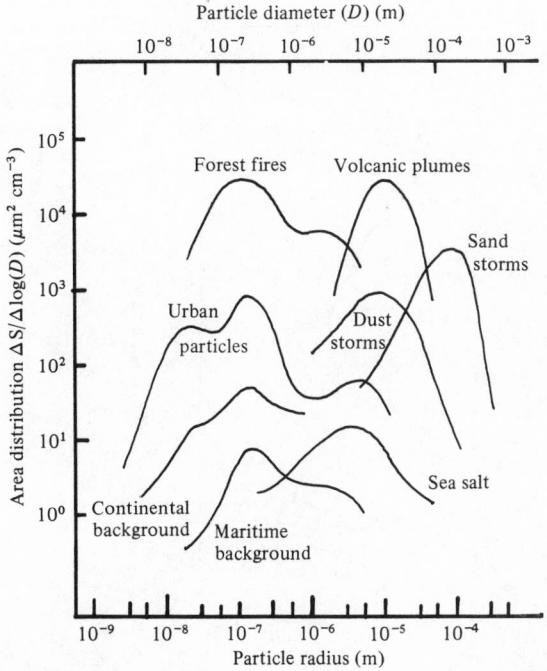

particles show enhanced areas in the giant particle size range. In volcanic particulate materials the giant particle area is much enhanced, due to the high velocities at which the particles are injected into the atmosphere. Naturally, such very large particles will have a limited lifetime in the atmosphere and are rapidly deposited onto the ground.

This brings us onto the question of removal of particulate materials from the atmosphere. The very small ones coagulate and the large ones can simply fall out. This means that particles of radius about 0.3 μm will appear to accumulate in air, by virtue of their long residence times (see Fig. 4.7). Particles can also be removed through incorporation into raindrops. Some particles, most probably those in the 0.1 μm radius range, and especially sulphate particles are effective condensation nuclei. These initiate the formation of cloud droplets that will be removed with any rain that falls. Smaller particles, i.e. those with radii less than 0.005 μm, rapidly diffuse to cloud droplets. Larger particles such as ammonium sulphate and sodium chloride aerosols' particles absorb water and are removed from the atmosphere by rainfall. Large, even insoluble ones, can simply be swept out of the atmosphere by falling raindrops. The close association between rainfall and removal of particles means that the maximum residence time in the lower troposphere is close to that of water (i.e. a little

Fig. 4.7. Residence times of particles in the troposphere (Jaenicke, 1978).

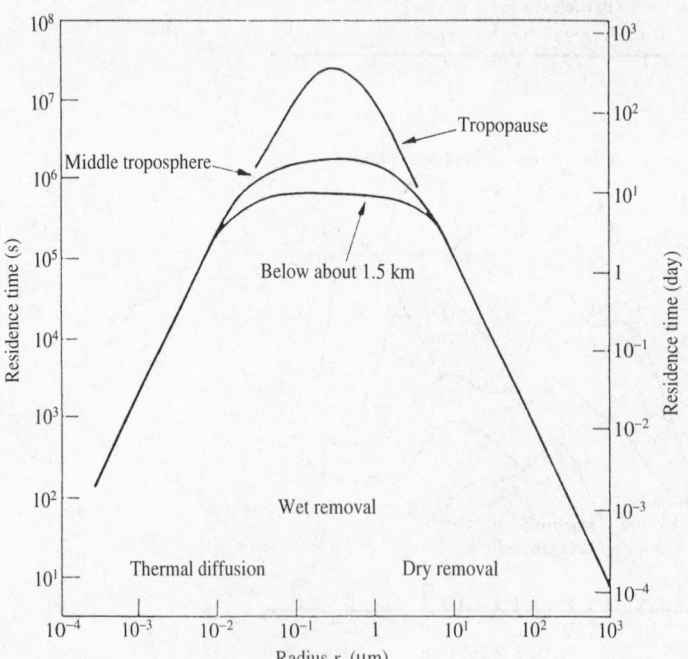

less than ten days). Often micrometre-sized suspended particles have to be brought near to the ground by turbulent transport and then undergo deposition through the still air boundary at the surface. Particulate materials are easily described as having deposition velocities in the same way as gases. Deposition velocity increases with size (see Fig. 4.8). In the larger size ranges it is close to the Stokes sedimentation velocity and then decreases with size. However, it increases when particles are very small because they diffuse to the ground rather more effectively.

4.3   Enrichment of trace components on particles

Now we will consider the composition of atmospheric particles in a little more detail. So far we have assumed that solid particles of dust or sea salt in the atmosphere resemble the material from which they were derived. This was delineated in Fig. 4.1, which shows the clear correspondence between the composition of crustal material and the suspended dust. Unfortunately, it gives an illusion of simplicity. In the atmosphere aerosols are not merely fragments of weathered, but chemically unaltered, crustal material. It is true that, when there is a great deal of dust suspended in the air, we can expect the agreement between crustal composition and aerosol composition to be good. However, when the aerosol is in low concentration or far from its source considerable modification of the composition is possible.

Fig. 4.8. The measured deposition velocity of particles over grass compared with the calculated rate of sedimentation (after Hidy (1973)).

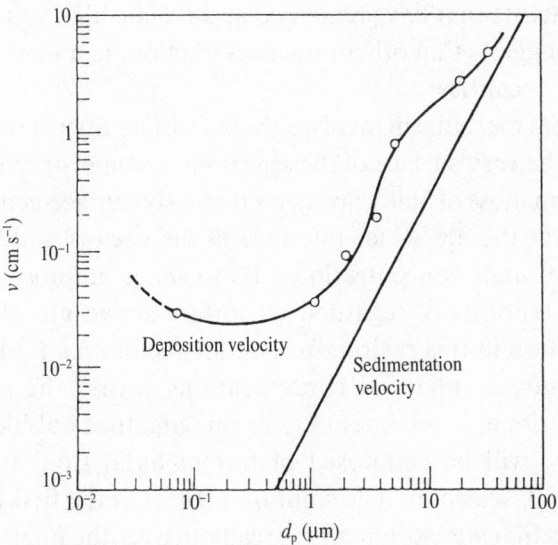

The composition of the aerosol above the sea, while largely sea salt, appears to have unexpected proportions of many seawater trace elements incorporated within it. Elements such as mercury, cadmium and selenium are frequently found in concentrations that are much too high compared with their relatively low concentrations in seawater. The degree of enrichment of trace elements in aerosols is given by an enrichment factor, EF. The concentration of a given element cX is related to a reference element, usually sodium when considering aerosols of marine origin, such that

$$EF(Na)_X = (c_X/c_{Na})_{air}/(c_X/c_{Na})_{sea} \tag{4.5}$$

or for continental aerosols it is normal to reference the element under consideration to aluminium so that

$$EF(Al)_X = (c_X/c_{Al})_{air}/(c_X/c_{Al})_{crust} \tag{4.6}$$

Crustal elements that have an EF of unity are those that lie on the line drawn in Fig. 4.1, i.e. they are exactly at the concentration that would be predicted from direct mechanical transfer into the atmosphere. Enrichment factors for many elements are far in excess of unity. In the marine atmosphere this is especially true of elements such as lead, iron and zinc. Human activities mobilise large amounts of these elements. This is illustrated by the fact that enrichment factors of many metals are higher over the North Atlantic ocean than above the remote Pacific Ocean. However, the high enrichment factors are not completely explained by long-range transport of anthropogenic pollutants. Ice cores in Antarctica have some enrichment factors that have remained at the same high values for over a century. This suggests that other processes fractionate materials onto aerosols in remote localities.

One possible enrichment mechanism involves the sea surface microlayer. This is the thin layer at the very surface of the sea. It has some properties that are quite different from those of bulk seawater. Obviously, surface-active species will accumulate at the air–water interface of the oceans and be found there at relatively high concentrations. However, a number of compounds that are not normally regarded as surface-active are also found at high concentration in this region. Accept, for the moment, that some materials are found at enhanced concentrations within the sea surface microlayer. This done, it becomes easy to imagine that bubbles, which generate aerosols, will be composed of this material from this region rather than the bulk seawater. The solution present in the first jet drop contains substances that were originally spread out over the interior surface of the bubble. This surface would have enhanced levels of surface-active components in much the same way as the sea surface. This mechanism for enrichment leads us to expect that materials that were enhanced in the microlayer would also be enhanced in the marine aerosol. There is some evidence that elements that are enriched in the microlayer can also be found to be enriched in marine aerosols (see Fig. 4.9).

Elements such as iron, zinc, lead and copper are not surface-active, but have been observed at enhanced concentrations in the sea surface microlayer. It has been argued that these metallic trace elements bind effectively to surface-active organic compounds. One might even go further and make a semi-quantitative prediction that the stronger the affinity between the metal and the organic compound, the stronger the enhancement in the microlayer. Hence large EF(Na) values for an element in the marine aerosol will be found for elements that form strong organic complexes. One of the difficulties is that we do not know the form of the organic compounds in the microlayer, so it is not possible to determine the binding strengths for the complexes of various metals. However, it is well known that metal complex formation follows a fairly regular pattern for a wide range of organic ligands. The transition metals found enriched in the marine aerosol form strong coordination compounds

Fig. 4.9. Enrichment factors, relative to sodium, for a number of elements in the surface microlayer and the marine atmospheric aerosol.

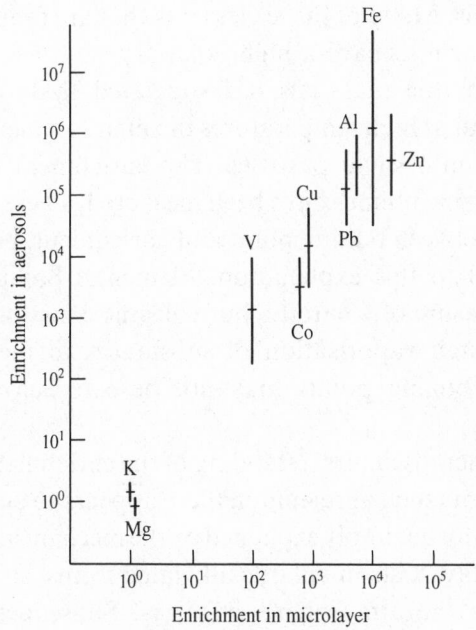

with many organic ligands. However, the agreement between these
notional association constants and enrichment factors is only partially
successful. It does not appear to explain all the elemental enrichment in
the atmosphere.

Complexation does not account for the enrichment of selenium, which
is sometimes observed in marine aerosols. It is hard to imagine a
non-metallic element such as selenium binding to carboxylic acids in the
same way as transition metals; yet selenium often shows a very pronounced
enrichment in the marine aerosol. In addition it should be noted that the
organic material occurs at very low concentration, even in the microlayer.
This means that there is rarely enough organic material to enhance the
concentration of anything but the trace elements of seawater. Furthermore
the sea surface microlayer, at least in its more visible form as slicks, is not
always present on the ocean.

Major sea salt elements, sodium, calcium, magnesium and potassium,
are found on large particles (3–7 μm). Elements likely to be of crustal
origin are enhanced because of the long-distance transport of continental
dusts, i.e. iron, aluminium and silicon, are associated with particles in the
1–3 μm size range. The highly enriched elements of the marine aerosol,
lead, zinc and copper, are found as very fine particles less than 0.5 μm
across. Such a small size range is often associated with high-temperature
processes. This suggests that the enrichment seen in marine aerosols takes
place during aerosol production processes rather than during those that
occur at the surface of the oceans. Many of those elements that are found
in fractionated onto aerosols seem to have a high volatility.

The enrichment of more volatile materials has suggested that the
fractionation process might occur at high temperatures through some sort
of distillation of the elements onto small particles. The enrichment of
elements on the aerosols in volcanic plumes have been measured. There is
often a reasonable agreement between boiling point and enrichment, but
many objections can be raised to this explanation. Elemental boiling
points may be a reasonable measure of volatility, but volcanic emissions
do not necessarily occur through vaporisation of substances in their
elemental state, so elemental boiling points may not be satisfactory
analogues.

At the moment there is no generalised understanding of the enrichment
process. In particular, for elements such as selenium, there appears to be a
widespread enrichment that cannot be simply explained by the mechanisms
discussed above. Selenium readily exists in volatile alkylated forms, so it
might well be transferred into the atmosphere as a gas. Subsequent

oxidation could convert it to less volatile compounds, which would adhere to particles. In this way biological alkylation of some elements could provide a mechanism for marine contribution to particles (as suggested in Table 2.3). Mercury and arsenic are methylated by micro-organisms with relative ease, and mercury may volatilise following reduction.

4.4   Absorption and chemistry on particle surfaces

Particulate material suspended in the atmosphere has a very high surface area per unit mass because of the small size of most of the particles. The area is often estimated at around $10^6 \, \mathrm{m^2 \, g^{-1}}$. Such a large area offers considerable opportunity for the absorption of molecules from the gas phase. This is particularly true, as we have just argued above, if they have a low volatility. It is generally thought that a substance whose vapour pressure is less than 10^{-6} Pa at ambient temperature will be largely adsorbed onto the atmospheric aerosol. This means that metals volatilised through volcanic or methylated through biological processes will ultimately end up bound to atmospheric aerosols as a metal or oxide.

Substances of moderate volatility can distribute themselves partly between the gas and particle phases. The process is idealised as

$$A_{(g)} + SP = A_{(SP)} \tag{R4.3}$$

where $A_{(g)}$ is the absorbing gas, SP the suspended particulate matter and $A_{(SP)}$ the gas absorbed on the particulate matter. This can be described in terms of a partition equilibrium constant:

$$K_p = [A_{(SP)}]/[A_{(g)}][SP] \tag{4.7}$$

The concentrations are conventionally in units of weight per unit volume of air. Some ratios have special significance, i.e. $[A_{(SP)}]/[SP]$ will represent the fraction of the suspended particulate matter that is the absorbed gas A (although this is reasonably exact only when the amount absorbed is small). The ratio $[A_{(SP)}]/([A_{(g)}] + [A_{(SP)}])$ is the fraction of A found on the suspended phase (see Fig. 4.10)

The way in which volatility is related to partition can be seen more clearly when we consider changes in the partition equilibrium constant with changes in temperature. A plot of the logarithm of the constant against reciprocal temperature tends to be linear (in rather the same way as it would be for the vapour pressure of liquids or solids where the inverse is known as the Antoine relationship) and obeys the equation

$$\log K_p = m_p/T + b_p \tag{4.8}$$

where m_p and b_p are constants. The quantity m_p is related to the enthalpy of desorption of the gas from the solid. The form of the equation reminds us that K_p will be small at high temperature and gases will tend to desorb from suspended particles (see Fig. 4.10). Less volatile compounds (those with high boiling points) have larger values of K_p, and often m_p, as determined from partition measurements.

The large surface-area-to-volume ratio of aerosols also increases the likelihood of surface reactions. A range of surfaces has been favoured as models for the surface of atmospheric aerosols: graphitic carbon, sulphuric acid and water. The latter two are of course liquids, but, as many particles in the atmosphere are acid sulphate droplets or hygroscopic water, these are reasonable representations of aerosol surfaces.

The rate of uptake of gases and vapours by particles is often given in terms of a sticking or accommodation coefficient. This is the ratio of the

Fig. 4.10. (*a*) Typical distributions of PAHs and biphenyl compounds between the gas and particulate phase in an urban environment. They are listed approximately in their order of elution from a gas chromatograph and plotted together with their boiling points, +. (*b*) Shows how the PAHs are built from multiple benzene rings.

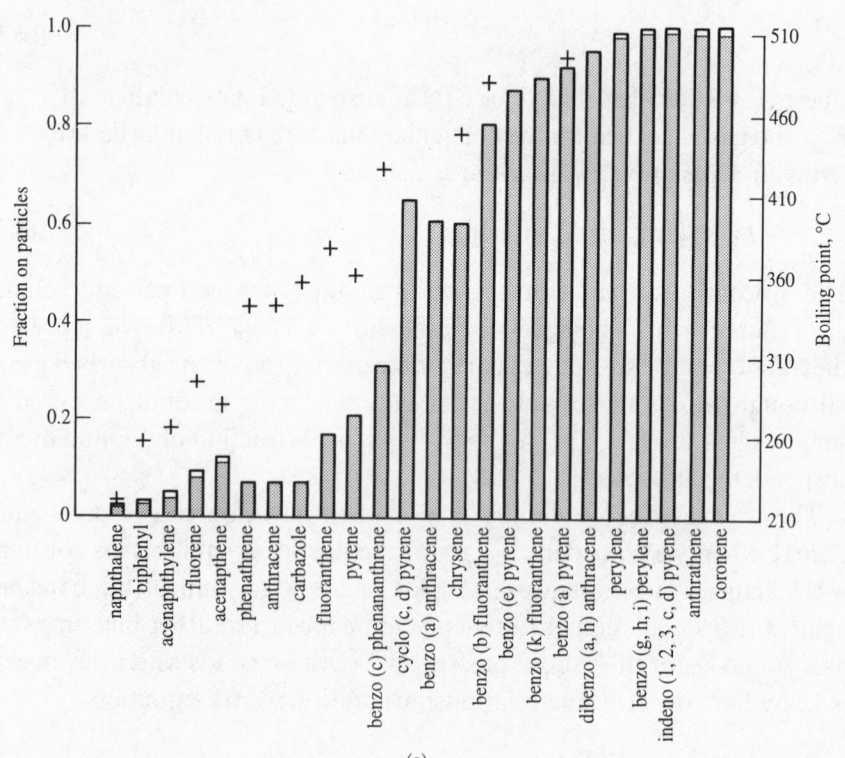

(a)

number of molecules absorbed by the surface to the number of collisions. Some typical accommodation coefficients are listed in Table 4.3. With sulphuric acid, even for a very reactive gas such as water vapour the molecule has to strike the sulphuric acid surface about five hundred times before it reacts. Sulphur dioxide, nitric acid or hydrogen peroxide are absorbed more readily by water surfaces.

Absorption of gases by graphitic carbon is well known. Sulphur dioxide, for example, is readily absorbed and in the presence of water is oxidised to sulphuric acid. Metallic oxides of manganese(II), vanadium(V), iron(III), copper(II) and aluminium are more effective than soot, with

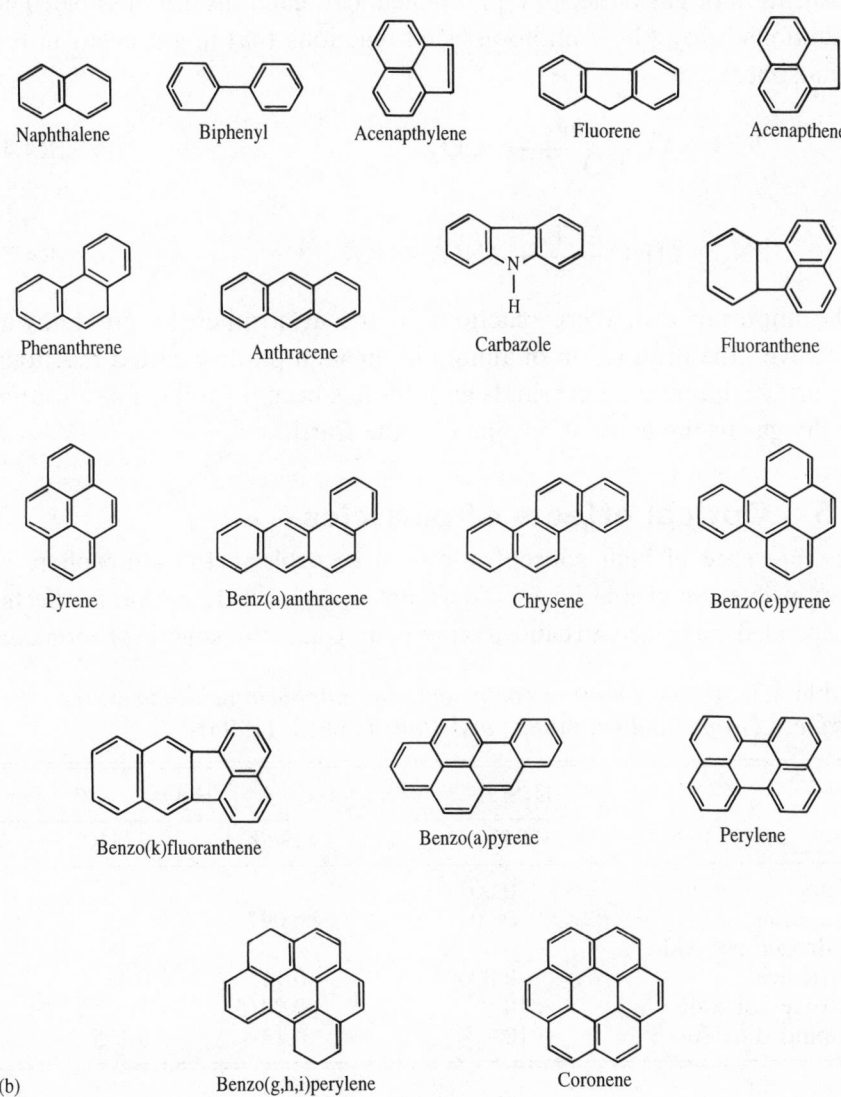

Naphthalene   Biphenyl   Acenapthylene   Fluorene   Acenapthene

Phenanthrene   Anthracene   Carbazole   Fluoranthene

Pyrene   Benz(a)anthracene   Chrysene   Benzo(e)pyrene

Benzo(k)fluoranthene   Benzo(a)pyrene   Perylene

(b)   Benzo(g,h,i)perylene   Coronene

mixtures of manganese and vanadium being especially effective. However it is not easy to assess the effectiveness of these processes in the global atmosphere, because the surface of the aerosol soon becomes saturated or poisoned. At this point further gas molecules will not be absorbed and oxidised unless there is some mechanism for 'cleaning' the surface. Liquid water would provide one way of achieving this within clouds or liquid aerosols. These complications mean that we do not know the extent to which oxidation processes on particle surfaces are removing gases, like sulphur dioxide, from the atmosphere.

Besides the possibility of thermal reactions on particle surfaces following absorption of gas molecules, photochemical reactions are possible. The equations below show photo-assisted reactions that might occur in the atmosphere:

$$2CO + O_2 \xrightarrow[\text{TiO}_2,\text{ZnO}]{hv} 2CO_2 \tag{R4.4}$$

$$2N_2 + 6H_2O \xrightarrow{hv,\text{TiO}_2} 4NH_3 + 3O_2 \tag{R4.5}$$

The importance of these reactions in the atmosphere is not known. However, the production of ammonia through photo-assisted reactions occurs on titanium desert sands and this has been postulated as a source of the gas in the early atmosphere of the Earth.

4.5   Optical effects of particles

The presence of high concentrations of aerosols in the atmosphere is responsible for visible hazes. There are even more remarkable effects. Suspended materials can cause a range of unusual atmospheric phenomena

Table 4.3. *Accommodation coefficients for sulphuric acid and water surfaces (from Baldwin (1982) and Ponche et al. (1993))*

	H_2SO_4	H_2O	
	300 K	298 K	273 K
Water	0.002		
Ammonia	0.001	0.097	
Hydrogen peroxide	0.0008		0.18
Nitric acid	0.00024	0.11	0.07
Nitrogen dioxide	$<10^{-6}$	0.0015	
Sulphur dioxide	$<10^{-6}$	0.13	0.115

such as blue moons, green suns and green flashes, or arcs about the sun or moon.

Our discussion will begin by looking at the way in which a single particle scatters light. The examination of a single particle is permissible because, since the particles are separated by a distance of more than 10–100 particle radii, the scattering of light is incoherent. Interparticle distances are generally greater than this in the atmosphere. There are two main theories of scattering: that of Rayleigh (dating from 1871) and the more complex theory of Mie (dating from 1908). In their usual form they treat scattering by transparent spherical particles. The Rayleigh Law for unpolarized light may be expressed as

$$I_0 = \frac{9\pi^2 V^2 (m^2 - 1)^2}{2d^2 \lambda^4 (m^2 + 2)} (1 + \cos^2 \theta) \tag{4.9}$$

where I_0 is the intensity of scattered radiation at distance d at unit incident illumination and angle θ from the particle, m is the refractive index of the particle and V its volume. The law implies that the scattered intensity will be proportional to r^6/λ^4. The difficulty with the Rayleigh theory is that it is only applicable to small particles. With visible light this is limited to particles of about 0.03 μm radius. This is somewhat smaller than the size that encompasses most of the atmospheric mass. However, one very general observation can be made from the theory. As the scattered radiation intensity is an inverse function of wavelength, the blue colour of the scattered light from the sky can be explained in terms of effective scattering at shorter wavelengths. Furthermore, the red colour of the sun as it sets is derived from light that has lost much of the blue region of the spectrum owing to the scattering that has taken place over a very long pathlength. In particular, the spectacular sunsets after volcanic eruptions or bushfires are well known and these arise from the higher than normal concentrations of very fine particulate matter in the atmosphere.

The more general Mie theory may be expressed by the equation

$$I_0 = \frac{\lambda^2}{8\pi^2 d^2} (i_1 + i_2) \tag{4.10}$$

where i_1 and i_2 are defined in terms of the coefficients of the electric and magnetic waves. Before the development of the modern computer it was extremely tedious to determine scattering from the Mie theory. The calculation is no longer so time-consuming, but the results are still complicated to display. The scattered light intensity is an intricate function of the scattering angle. Scattering patterns vary with the wavelength of the light so colours can be perceived by an observer. The total scattering by a particle in all directions is given by

$$S = \frac{\lambda^2}{2\pi} \sum_{n=1}^{\infty} (2n + 1)(|a_n|^2 + |b_n|^2)$$
(4.11)

where a_n and b_n are the coefficients of the electric and magnetic waves. The extinction efficiency, Q_e, is a useful parameter that can be defined in terms of the total scattering:

$$Q_e = S/(\pi r^2)$$
(4.12)

Fig. 4.11. The extinction efficiency for a 0.3 μm radius particle with the refractive index of water (1.33) and sulphur (2.0) as a function of wavelength. The graphs can be used for different sized particles by computing the quantity α and using the scale at the top of the graph (after Sinclair (1950)). The quantity α is defined as $2\pi r/\lambda$.

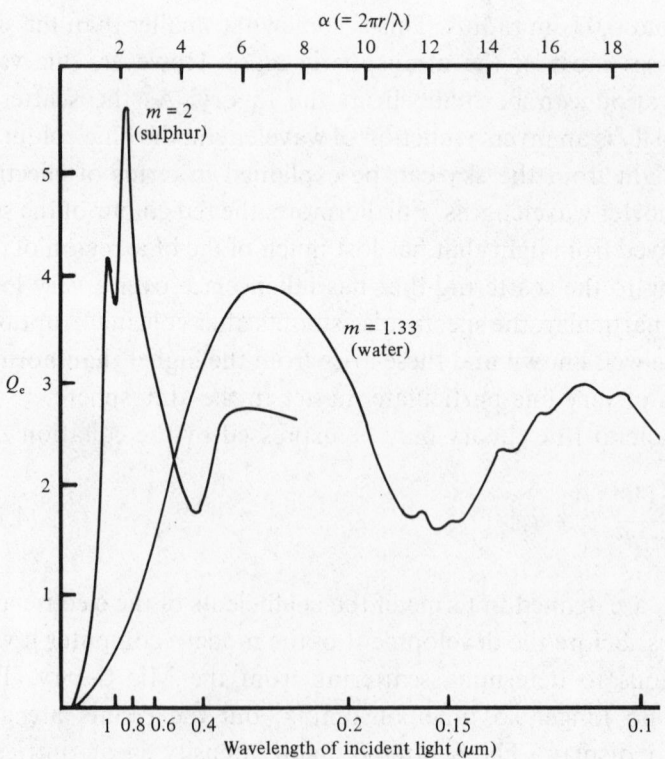

The extinction efficiency is the ratio of the apparent scattering cross-sectional area of the particle to the geometrical cross-sectional area. This value Q_e is a complex function of the radius of the particle with respect to the wavelength of the incident light and the refractive index of the particle (Fig. 4.11).

Q_e can be incorporated into the well-known Beer–Lambert Law, which describes absorption of light on a macroscopic level. When light passes through an absorbing medium the intensity declines exponentially:

$$I = I_0 \exp(-bL) \tag{4.13}$$

where I_0 is the initial light intensity and I the intensity after the light has passed along path length L through the medium and b is the extinction coefficient. This may be thought of as the sum of extinction due to both

Fig. 4.12. Extinction coefficient as a function of relative humidity for ammonium sulphate (mean radius 0.2 μm) and ammonium nitrate (mean radius 0.6 μm) particles. The salts are present at $1 \, \mu\text{m m}^{-3}$. The lines marked (a) are for a monodisperse aerosol and the lines (b) for a polydisperse one. The curves stop at low humidities, at which the crystal dries out (data from Tang et al., (1981)). The data points in open circles are on a different scale and represent measured visibility at various relative humidities when the ammonium sulphate concentration was $20 \, \mu\text{m m}^{-3}$.

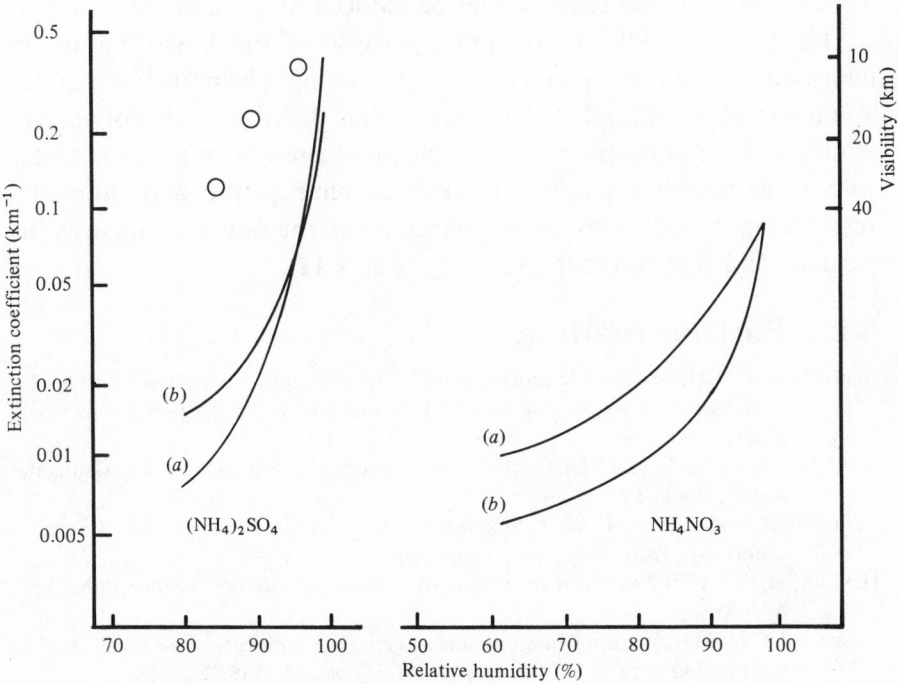

absorption (coefficient b_a) and scattering (coefficient b_s). It is possible to determine the value of b_s from the scattering coefficient:

$$b_s = \pi r^2 N Q_e \tag{4.14}$$

where N is the number of scattering particles per unit volume. Normally the particles are not of uniform radius so the value of b_s would have to be determined by summing over the range of particle sizes. The scattering coefficient (and also the extinction coefficient) takes the units of reciprocal length.

The absorption of light can be related to visibility by realising that visibility is limited by difficulties in perceiving the contrast difference between two distant objects. The decrease in contrast is responsible for the haziness we notice in particle-laden atmospheres. When contrast differences drop to about 2% they are rendered invisible in a practical sense. Contrast differences (C) follow the Beer Law in much the same way as light intensity, so

$$C = C_0 \exp(-bL) \tag{4.15}$$

Normally a high-contrast object would be taken as the visible target. If this target is black and white then C_0 is unity, by definition, so that rearranging the equation gives $L = 3.9/b$. Thus, at a b value of $4 \times 10^{-3}\,\mathrm{m}^{-1}$, visual range would be reduced to a kilometre.

The theory is difficult to apply, because of the large amount of information required to predict the Mie scattering. The refractive index as a function of wavelength and particle radius distribution is not easy to establish in the atmosphere. This is partly because these quantities vary with meteorological parameters such as temperature and humidity. Increases in humidity can cause an increase in the radius of hygroscopic particles and thus lower the visibility (Fig. 4.12).

4.6   Further reading

Buat-Menard, P. (1983) Particle geochemistry in the atmosphere and oceans, in *Air Sea Exchange of Gases and Particles*, ed P. S. Liss and W. G. N. Slinn, pp. 455–532, Reidel, Dordrecht.

Götz, G., Mészáros, E. and Vali, G. (1991) *Atmospheric Particles and Nuclei*, Ákádemiai Kiadó, Budapest.

Leslie, G. B. and Lunau, F. W. (1992) *Indoor Air Pollution Problems and Priorities*, Cambridge University Press, Cambridge.

Hesketh, H. E. (1977) *Fine Particles in Gaseous Media*. Ann Arbor Science Publishers, Ann Arbor.

Pankow, J. F. (1991) Common *y*-intercept and single compound regressions of gas–particle partitioning data vs $1/T$. *Atmospheric Environment*, A**25**, 2229–39.

Prospero, J. M., Charlson, R. J., Mohnen, V., Jaenicke, R., Delaney, A. C., Moyers, J.,
Zoller, W. and Rahn, K. (1983) The atmospheric aerosol system: an overview,
Reviews in Geophysics and Space Physics, **21**, 1607–29.

5

○ ○

The chemistry of aqueous systems

This chapter considers the chemistry of liquid water in the atmosphere. Together with rainwater, we often have to account for the water included within cloud droplets, dew droplets and even the oceans.

5.1   Cloud physics

The physics of cloud droplet formation affects the chemistry of precipitation. Last century it was realised that water vapour does not necessarily condense in clean air even when the vapour pressure of water is much greater than that required to form liquid water. This high degree of supersaturation arises because the equilibrium vapour pressure over small droplets is much larger than that over plane surfaces (p_∞). The relationship between the radius of a water droplet and the partial pressure of water vapour (p_r) above it is given by

$$\log_e (p_r/p_\infty) = 2\gamma M/(\rho_L R T r) \tag{5.1}$$

where γ, ρ_L and M are the surface tension, density and molecular weight of water. In Fig. 5.1 are shown the vapour pressure and percentage saturation above a small water droplet. The pronounced effect of the presence of salts in the water is also shown. Salts lower the water vapour pressure considerably, and it is in this way that they act as condensation nuclei. Subsequently, their presence contributes to the composition of precipitation.

The water vapour removed by condensation would soon lead to the vapour pressure falling below the saturation level and droplet growth would cease. In a cloud the condition of supersaturation can be maintained by the cooling of ascending air. Colder air becomes saturated at a lower water vapour content, hence the supersaturation can continue and so can droplet growth. High up in a cloud, temperatures may drop well below the freezing point of water, so ice, snow, rain and hail can form. The initial rate of growth of cloud droplets is quite rapid, but after the radius reaches some 20 μm or so, condensation becomes an inefficient mechanism for growth. Once droplets are this big they can coalesce through collisions. In Fig. 5.2 are shown idealised distributions of cloud and rain droplets. The axes are similar to those used for particles in Section 4.2. Note that there are pronounced maxima in the curves, which mark the size that occurs with the greatest frequency. Size ranges covered by cloud and rain droplets are very different, with heavy rain having higher frequencies at the largest sizes.

5.2 The solubility of gases

An important influence on rainfall chemistry is the dissolution of atmospheric trace gases into suspended droplets. The solubility of gases in

Fig. 5.1. Relative humidity of air in equilibrium with droplets of various size containing different amounts of salt. Very small droplets of pure water would require the air to be supersaturated in order for condensation to occur. However, the presence of even a small amount of salt as condensation nuclei lowers the equilibrium relative humidity remarkably.

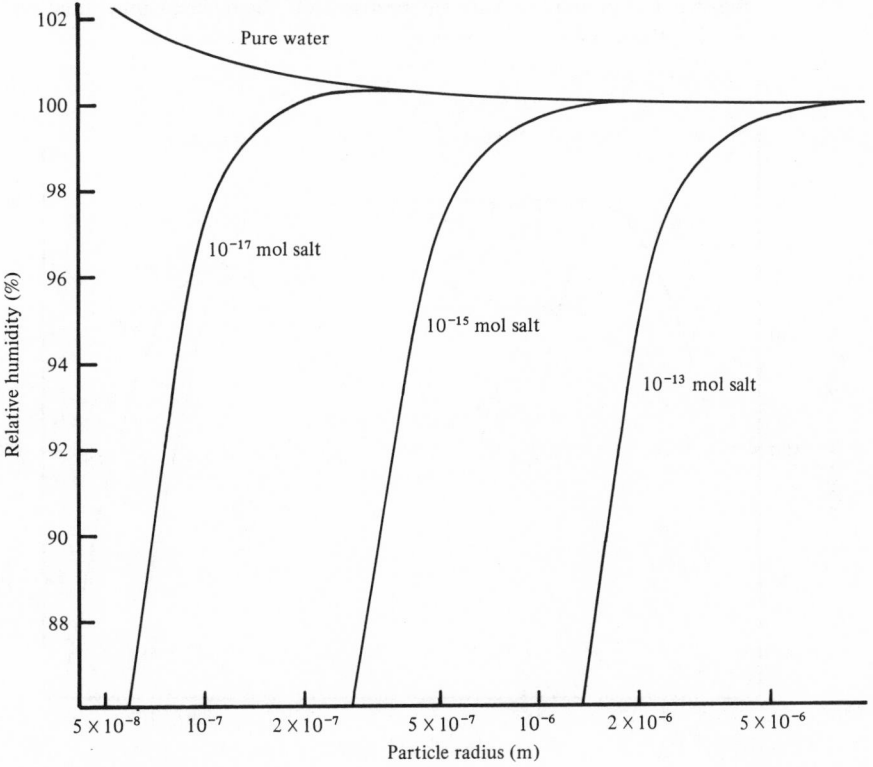

water is conveniently described by a relationship known as Henry's Law, which states that, at equilibrium, the partial pressure of a gas above a solution of it is proportional to the concentration of the gas in the solution. However, in much atmospheric chemistry it is useful to imagine the relationship between the gas and liquid phase concentrations in terms of an equilibrium of the type

$$A_{(g)} \rightleftharpoons A_{(aq)} \tag{R5.1}$$

taking p_A and $[A_{(aq)}]$ to represent the concentration of substance A in the gas and liquid phase. Writing a constant of Henry's Law K_H as the equilibrium constant for this reaction using pressure to describe the concentration of A in the gas phase, we have

$$K_H = [A_{(aq)}]/p_A \tag{5.2}$$

If we take the atmosphere as the unit of pressure and $mol\,l^{-1}$ as the unit of concentration, then the constant of Henry's Law will have the units $mol\,l^{-1}\,atm^{-1}$. Typical values of the constant are given in Table 5.1. The larger the values of this constant the more soluble the gas, thus a material

Fig. 5.2. Cloud and rain droplet spectra. Note the different units used for plotting the cloud and rain droplets.

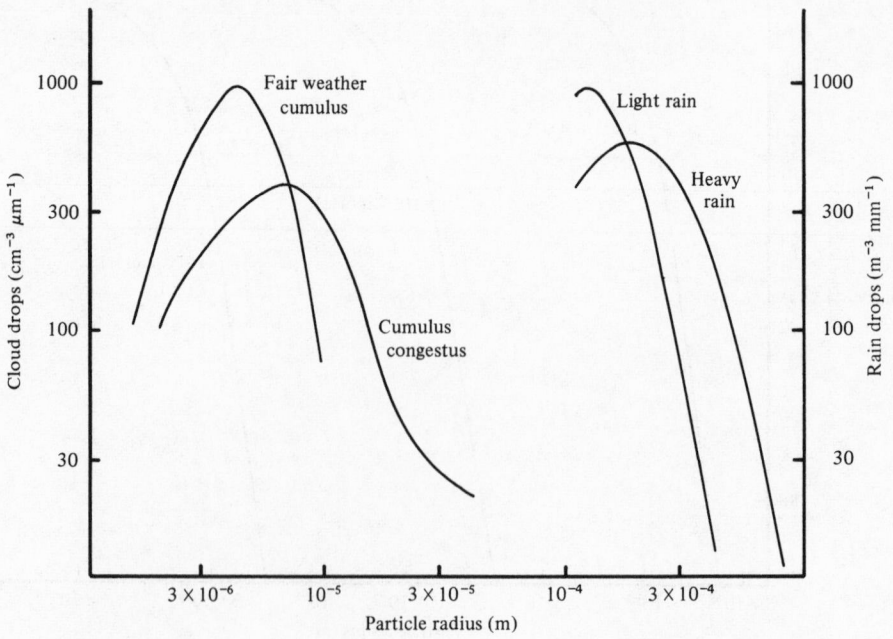

like hydrogen peroxide is highly soluble and large amounts can dissolve in cloud and rain droplets.

The situation with gases that react in water is a little more complicated. The constant of Henry's Law accounts for only simple dissolution and not the subsequent hydrolysis. For instance, formaldehyde dissolves and hydrolyses to methylene glycol according to the equations

$$HCHO_{(g)} = HCHO_{(aq)} \qquad \text{(R5.2)}$$

$$HCHO_{(aq)} + H_2O_{(l)} = H_2C(OH)_{2(aq)} \qquad \text{(R5.3)}$$

This hydrolysis means that the apparent solubility of formaldehyde in water is greater than that expected from the constant of Henry's Law. The total amount of formaldehyde dissolved, T(HCHO), in solution via reactions (R5.2) and (R5.3) would be

$$T(HCHO) = [HCHO_{(aq)}] + [H_2C(OH)_{2(aq)}] \qquad \text{(5.3)}$$

The concentration of methylene glycol will be related to the aqueous formaldehyde concentration by the law of mass action:

Table 5.1. *Henry's law constants for some atmospheric trace gases at* $15\,^\circ C$

Gas	K_H (mol l^{-1} atm^{-1})
Hydrogen peroxide	2×10^5
Dieldrin	5800
Lindane	2230
Ammonia	90
Aldrin	85
DDT	28
Sulphur dioxide	5.4
Formaldehyde	1.7
Mercury	0.093
Carbon dioxide	0.045
Acetylene	0.05
Nitrous oxide	0.034
Ozone	0.02
Nitric oxide	0.0023
Methane	0.0017
Oxygen	0.0015
Nitrogen	0.001
Carbon monoxide	0.001

$$K = [H_2C(OH)_{2(aq)}]/[HCHO_{(aq)}] \tag{5.4}$$

so

$$T(HCHO) = [HCHO_{(aq)}] + K[HCHO_{(aq)}] \tag{5.5}$$

and, as we know $[HCHO_{(aq)}]$ from Henry's Law, the equation above may be rewritten as

$$T(HCHO) = K_H p(HCHO)(1 + K) \tag{5.6}$$

In the case of formaldehyde K is about 2000, i.e. the gas is readily hydrolysed by water, so most of it will be found in the aqueous solution as methylene glycol rather than formaldehyde. This makes formaldehyde a soluble gas. The constant of Henry's Law is about $1.7\,mol\,l^1\,atm^{-1}$. At equilibrium with an atmosphere with $p(HCHO)$ at $10^{-9}\,atm$, we would predict a total concentration of formaldehyde-derived carbon of $3.4 \times 10^{-6}\,mol\,l^{-1}$.

The dissolution of formaldehyde is a simple case, because many gases undergo more complex hydration reactions in water. A particularly important set of reactions influence the pH of rainwater: namely the dissolution of carbon dioxide, sulphur dioxide and ammonia. These dissolve according to the equations

$$CO_{2(g)} + H_2O_{(l)} = H_2CO_{3(aq)} \tag{R5.4}$$

$$H_2CO_3 = H^+ + HCO_3^- \tag{R5.5}$$

$$HCO_3^- = H^+ + CO_3^{2-} \tag{R5.6}$$

Reactions (R5.5) and (R5.6) are in aqueous solution, but the state of the species is not specified. For simplicity the subscript (aq) will be omitted through the rest of this chapter unless a phase change or other special feature is involved.

$$SO_{2(g)} + H_2O_{(l)} = H_2SO_{3(aq)} \tag{R5.7}$$

$$H_2SO_3 = H^+ + HSO_3^- \tag{R5.8}$$

$$HSO_3^- = H^+ + SO_3^{2-} \tag{R5.9}$$

$$NH_{3(g)} + H_2O_{(l)} = NH_4OH_{(aq)} \tag{R5.10}$$

$$NH_4OH = NH_4^+ + OH^- \tag{R5.11}$$

The constant of Henry's Law and the equilibrium constants for these reactions are given as K_H, K' and K'' respectively in Table 5.2. It can be seen in the cases of both carbon dioxide and sulphur dioxide that the second dissociation constants are very much smaller than the first. This means that the second dissociation may be neglected at acidic pH values. Thus the pH of a droplet of water in equilibrium with atmospheric carbon dioxide can be determined by combining two equilibrium constant equations, one governing the dissolution and the other the first step in the dissociation. The concentration of HCO_3^- will be

$$[HCO_3^-] = K_H K' p(CO_2)/[H^+] \tag{5.7}$$

If the only source of hydrogen ions in the system is from the dissociation of carbon dioxide, then $[HCO_3^-] = [H^+]$ so that

$$[HCO_3^-] = [H^+] = (K_H K' p(CO_2))^{1/2} \tag{5.8}$$

Substituting the appropriate values of the equilibrium constants and using a carbon dioxide partial pressure of 360 ppm, i.e. 3.6×10^{-4} atm will yield a hydrogen ion concentration of $2.5 \times 10^{-6}\,mol\,l^{-1}$ or a pH of 5.6.

   In remote regions, the pH of very pure rainfall can be close to this, although even there trace amounts of other compounds can affect the acidity. If a small amount of sulphur dioxide is present in the air at a concentration of 5×10^{-9} atm (which is not unreasonable over continental land masses), then we can write an equation for the pH of a solution in equilibrium with sulphur dioxide that is analogous to the one just used for carbon dioxide:

$$[HSO_3^-] = [H^+] = (K_H K' p(SO_2))^{1/2} \tag{5.9}$$

and using the appropriate constants from Table 5.2 gives a pH of 4.6. Thus, even at low concentrations, sulphur dioxide affects the pH. Thus, although carbon dioxide is found at much higher atmospheric concentra-

Table 5.2. *Henry's law constants and equilibrium constants for atmospheric gases that undergo hydrolysis*

Gas	K_H $(mol\,l^{-1}\,atm^{-1})$	K' $(mol\,l^{-1})$	K'' $(mol\,l^{-1})$	(°C)
Ammonia	90	1.6×10^{-5}		15
Pyruvic acid	3.1×10^5	3.4×10^{-3}		25
Formic acid	5.2×10^3	1.8×10^{-4}		25
Acetic acid	5.2×10^3	1.75×10^{-5}		25
Hydrogen fluoride	1.4×10^4	6.7×10^{-4}		25
Sulphur dioxide	$5.42.7 \times 10^{-2}$		10^{-7}	15
Carbon dioxide	0.045	3.8×10^{-7}	3.7×10^{-11}	15

tions than is sulphur dioxide, the high solubility and dissociation constants for sulphur dioxide make it more effective at acidifying water droplets than is carbon dioxide. Sulphur dioxide can have an even more dramatic effect on pH if we remember that it can be oxidised to sulphuric acid comparatively easily. This oxidation yields a further proton. One should also observe that the equations describing the dissolution of an acidic gas such as sulphur dioxide show that the presence of acids already in solution will depress the solubility of the acidic gas. Alkalis will enhance it.

Ammonia is the only common alkaline gas in the atmosphere. It will neutralise dissolved acids, in particular sulphuric acid, which means that ammonium sulphate is frequently found as a salt in the atmosphere. For example, let us assume a system of water in equilibrium with the following three pressures of CO_2, SO_2 and NH_3: $p(CO_2) = 3.6 \times 10^{-4}$ atm, $p(SO_2) = 5 \times 10^{-9}$ atm and $p(NH_3) = 10^{-9}$ atm. We can solve for pH as we have done for a single gas, except that it is a little more complex because the solution is not particularly acidic and the second dissociation constant of sulphurous acid becomes important. The pH obtained is about 5.8, showing that ammonia has an effect on pH even at very low concentrations.

Although we often think of sulphur dioxide as a highly soluble gas, this does not mean that all of it dissolves in the liquid phase of a system of cloud droplets suspended in air. The ratio of the volume of water to the volume of air is so low (less than 10^{-6}) that a gas has to be very soluble to be found predominantly within the liquid phase. We can calculate the critical value of K_H at which a gas would be equally distributed between each phase, i.e.

$$TA_{(g)} = TA_{(aq)} \tag{5.10}$$

where $TA_{(g)}$ and $TA_{(aq)}$ are the total amounts of the gas, A, in the gas and aqueous phase in a cubic metre of air. The amount $TA_{(aq)}$ can be expressed in concentration terms:

$$TA_{(aq)} = [A_{(aq)}]V_w \tag{5.11}$$

where V_w is the volume of liquid water in a cubic metre. Henry's Law allows the concentration to be rewritten in terms of pressure:

$$TA_{(aq)} = K_H p(A)V_w \tag{5.12}$$

which has the units mol (for a unit volume of air). We can convert the amount in the gas phase into pressure units also by dividing by the volume of a mole at one atmosphere pressure (i.e. about $0.0245\,m^3$) such that

$$TA_{(g)} = p(A)/0.0245 = K_H p(A)V_w \tag{5.13}$$

which means that

$$K_H = 1/(0.0245 V_w) \tag{5.14}$$

This critical constant of Henry's Law will be about $4 \times 10^4 \, \text{mol} \, l^{-1} \, \text{atm}^{-1}$ if $V_w > 0.001 \, l$ (i.e. about a gram of liquid water per cubic metre in very heavily water-laden clouds).

We can see from Tables 5.1 and 5.2 that only hydrogen peroxide and pyruvic acid are soluble enough to partition predominantly into the liquid phase without further reaction. However, if gases dissociate or react in water then their solubility can be much enhanced. Sulphur dioxide, for example will partition into the liquid phase when the pH is above 5.5. This is the critical pH above which all the different equilibrium S(IV) species in solution will exceed those in the gas phase in a cloud laden with liquid water at $1 \, \text{g} \, \text{m}^{-3}$.

Strong acids are also found as trace gases in the atmosphere. These include HCl, HBr and HNO_3, which dissolve directly as the dissociated form:

$$HCl_{(g)} = H^+_{(aq)} + Cl^-_{(aq)} \tag{R5.12}$$

The constant of Henry's Law needs to be written to account for this direct dissociation as

$$K_H = [H^+_{(aq)}][Cl^-_{(aq)}]/p(HCl) \tag{5.15}$$

The equilibrium is shifted so far to the right that these gases can dissolve effectively in the tiniest volumes of water associated with aerosol particles. This means that they can achieve such high concentrations (i.e. in the $\text{mol} \, l^{-1}$ range) that concentration units are no longer an adequate description of the equilibrium relationship and it is usual to write it in terms of activity:

$$K_H = a(H^+_{(aq)})a(Cl^-_{(aq)})/p(HCl) \tag{5.16}$$

The constants of Henry's Law for these extremely soluble gases are given in Table 5.3.

Gases that transfer to the liquid phase effectively are efficiently removed from the atmosphere by rainfall. Thus, the conversion of insoluble trace gases into highly soluble ones represents a mechanism for their removal from the atmosphere.

5.3   The transfer of gases to liquids

While the equilibrium situation, described in the section above, is an important one, not all gases found dissolved in water are in equilibrium with air. Gases are sometimes at high concentration in the atmosphere and at extremely low concentrations in water bodies. Because of this there can be a flux of these gases to the oceans, for instance (as discussed in Section 2.4).

The flux of gases across an air–liquid boundary is usually described in terms of a two-film model. This assumes that there are thin boundary layers on either side of the gas–liquid interface and that transfer through these layers is controlled by diffusion. As diffusion is a slow process compared with turbulent transport, transfer through the still boundary layers limits the flux of gases to bodies of water such as the ocean. As mentioned in Section 2.4, the total resistance to transport is the sum of the individual resistances, so in gas–liquid transfer it will be the sum of the individual resistances of the two layers. For most important atmospheric gases, one of these two resistances is very much greater that the other. This makes it possible to describe the transport of gases across an air–liquid interface as either gas-phase- or liquid-phase-controlled. For highly soluble gases it is generally the gas phase that controls the transfer, while the transfer of less soluble gases is controlled by liquid-phase resistance.

In gas transfer studies the flux (F) across a boundary layer at the air–water interface is given by the equation

$$F = k\,\Delta c \tag{5.17}$$

where k is the transfer velocity and has units $m\,s^{-1}$ in a way which is analogous to the deposition velocity V_g discussed in Section 2.4 and Δc is the difference in concentration between the gas and liquid phases. In a simple case in which a gas is transferring from the gas to liquid phase the

Table 5.3. *Henry's law constants for atmospheric gases that are strong electrolytes dissolving according to the equilibrium* $HX_{(g)} = H^+_{(aq)} + X^-_{(aq)}$ *and defined as* $K_H = a(H^+_{(aq)})a(X^-_{(aq)})/p(HX)$

Gas	K_H at 25 °C $(mol^2\,l^{-2}\,atm^{-1})$
Hydrogen chloride	2.04×10^6
Hydrogen bromide	1.32×10^9
Hydrogen iodide	2.50×10^9
Nitric acid	2.45×10^6
Methanesulphonic acid	6.5×10^{13}

concentration in water might be low, almost zero. Under these conditions Δc will be the gas phase concentration and the transfer coefficient will control the concentration gradient set by this concentration difference. The exchange constant is the reciprocal of the resistance r. It can be obtained from the diffusion coefficient of the gas, D, in the boundary layer under consideration, and the thickness of the boundary layer, z, i.e. $k = 1/r = D/z$. As resistances can be summed, exchange constants must be summed as reciprocals, i.e. $1/K = 1/k_1 + 1/k_2 \ldots$, where K is the exchange constant of a whole system of boundary layers with individual exchange constants $k_1, k_2 \ldots$ etc.

Water vapour transfer to the oceans is controlled by gas-phase boundary layer diffusion because it does not have to be transferred through any liquid-phase boundary layer. There is another way in which transfer into liquids can be fast. If a gas reacts rapidly in the liquid then the resistance in the liquid-phase boundary layer may be lowered, so that transfer becomes gas-phase-controlled. This is true of a gas such as sulphur dioxide, which rapidly hydrolyses to bisulphite and sulphite ions in water. Thus, sulphur dioxide dissolution into the oceans is under gas-phase control and this is rapid, i.e. about 0.5–$1 \, cm \, s^{-1}$. This hydrolysis is only significant for sulphur dioxide at pH values higher than 3, so in very acid solutions the dissolution will remain under liquid-phase control.

Dissolution of gases in falling droplets involves some mixing processes but can still be understood by considering the process in two steps. The first, involving transfer of the gas to the surface of the droplet from the bulk atmosphere, and the second mixing it within the droplet. The gas-phase transport processes are usually considered to be fast, so the rate of transfer is limited by mixing within the droplet. If the gas is transferred through the gas-phase boundary layer quite rapidly then the surface of the droplet can reach equilibrium with the gas phase. More gas can only dissolve in the droplet after the dissolved gas has mixed inwards from the droplet surface towards the centre. If the drop is falling, this may occur through convective stirring. If the liquid is stagnant, then it has to occur through the much slower diffusive processes, although these processes are still quite rapid in comparison with the lifetime of most droplets. The equation for diffusive transport within a sphere is

$$\frac{M_t}{M_\infty} = 1 - \frac{6}{\pi} \sum_{n=1}^{\infty} \frac{1}{n^2} \exp\left(-\frac{Dn^2\pi^2 t^2}{r^2} \right) \tag{5.18}$$

where M_t and M_∞ are the mass of the substance at time t and at equilibrium and n is an integer. Substituting typical values of $50 \, \mu m$ for the radius and a diffusion coefficient of $10^{-9}\,m^2\,s^{-1}$ (which is typical for many dissolved gases) suggests that 50% saturation is achieved in 0.3 s. This suggests that the process of equilibration of droplets with atmospheric gases is rapid. Diffusion in large rain droplets may take longer, but they may be stirred by convection as they fall.

Chemical reactions within the droplets can be very fast, but they are nonetheless often the rate-limiting steps in many processes, such as the oxidation of S(IV), that are of interest in the atmosphere.

5.4   The composition of precipitation

Insoluble solids are frequently found in rainfall. This represents an inert residue, although the surface may be active enough to promote some reactions (see Section 4.4). Soluble material plays the major role in the chemistry of precipitation. About 30% of the atmospheric aerosol greater than 4.5 µm in diameter is soluble enough to dissolve in atmospheric water. The fraction that is soluble increases with decreasing aerosol size. The major soluble ions are SO_4^{2-}, NO_3^- and NH_4^+. In the marine atmosphere the ions of seawater are also present in significant amounts. Marine carbonate and bicarbonate ions represent a limited source of buffering against the acids that form so readily in atmospheric solutions. Alkaline continental dusts will also help to reduce the acidity of rainfall, but their effectiveness will depend on incorporation into the raindrop and then subsequent dissolution. Dust is also a potential source of dissolved trace metal ions.

Bubble-bursting processes in the oceans inject vast quantities of sea salt aerosol into the atmosphere. While the greater part of this is returned directly to the oceans, large amounts are carried over the continents, especially during storms, and deposited with rainfall. The composition of rain in coastal regions is dominated by the presence of sodium chloride, the concentration of which decreases exponentially with distance from the coast, halving over distances of about 20 km. Beyond 100 km the concentrations level off at about $30\,\mu mol\,l^{-1}$ ($1\,mg(Cl)\,l^{-1}$).

This major influence of the sea on the composition of seawater is embodied in Dean's Rule, which states that an artificial rainfall can be made by diluting 1.5 ml of seawater to a litre of distilled water. This marine influence on the composition of rainfall is so strong that rainfall composition is often thought of in terms of the way it deviates from that of dilute seawater. It is common to consider how much the Cl/Na ratio varies from that found in seawater. If it is very much less than the seawater value then it can be taken as a hint that there has been a loss of chloride to the gas phase, perhaps as HCl. The term 'excess sulphate' or 'non-sea-salt sulphate' is frequently used to describe the sulphate present in rainfall. This is the excess over that expected from the SO_4/Na (or perhaps SO_4/Mg) ratio in seawater. This assumes that all the sodium and magnesium comes from sea salt, but in some regions there may be other sources.

Besides the dissolution of solids, gases represent the other important source of solutes for rainfall. There are few alkaline trace gases in the atmosphere. Ammonia is the most important, but it is rare that there is enough ammonia to neutralise the free acidity. If we consider aerosol particles, NH_4HSO_4 is more frequently found than the $(NH_4)_2SO_4$ that might be expected after complete neutralisation. There will be a few amines that will also serve to neutralise acids, but they are at even lower concentrations than is ammonia.

It is difficult to give general figures for the composition of precipitation because it varies so greatly. However, some examples of analyses are given in Table 5.4. The analyses from Oregon show a strong maritime influence with high concentrations of both sodium and chloride. By contrast the Pyrenees and Alaska have lower concentrations and the Pyrenean rainfall has a high sulphate-to-chloride ratio. Despite its remote location, rainfall there is still influenced by the large amounts of industrial sulphur emitted from Europe. The Alaskan samples also have sulphate concentrations enhanced through the addition of sulphuric acid produced in the atmosphere even at this remote location. The rainfall from the Pyrenees also seems to show high calcium concentrations that could come from the incorporation of dust.

Table 5.4. *Composition of rain* $(\mu mol\, l^{-1})$ *from various sites*

	Alaska[a]	Oregon[a]	Pyrenees[b]	Marine[c]
Calcium	0.25	0.5	47	11
Magnesium	1	2.5	4	55
Potassium	1	4	9	10
Sodium	9	51	23	482
Ammonium	2.4	3	22	1
Sulphate	6.7	4.1	25	29
Nitrate	0.8	2.9	18	3
Chloride	7	66	20	560

[a]Bormann *et al.* (1989).
[b]Camarero and Catalan (1993).
[c]Hypothetical composition of rainwater that is simply a thousand-fold dilution of seawater.

Another important point to note is that the sulphate and nitrate concentrations also contribute to a large fraction of the acidity. These will form from the dissolution of sulphur dioxide and its subsequent oxidation to sulphuric acid or the direct dissolution of nitric acid. In remote and unpolluted areas formic and acetic acid can make the most important contribution. These two acids make up the major fraction of the dissolved organic material found in rain. There are small amounts of other organic acids such as glyoxylic and pyruvic acid. In remote areas methanesulphonic acid, derived from the oxidation of DMS, can also make a significant contribution to the rainfall acidity. The addition of these further acids is responsible for the pH being often lower than the 5.6 estimated from equilibration with carbon dioxide.

Rainfall composition is highly variable. It is often given in terms of an annual average. However there is pronounced seasonal variation that sometimes shows a maximum for species such as nitrate in the spring. This may arise through increasing photochemical activity at this time of year. There are strong differences between falls of rain. Obviously this can depend on the origin of the air mass. If the rain has come directly from the sea then high concentrations of sodium and chloride are expected. If, on the other hand, it has come across an industrialised region then sulphate is expected to be high. Even within rainfall events there are sharp changes in concentration. The early parts of a rain event usually show the highest solute concentrations, because soluble trace components have yet to be removed from the air mass. Rainfall rate also affects concentrations, with heavier rain causing greater dilution of dissolved solutes.

Other forms of precipitation such as snow or dew may have different compositions from rain. Snow is a more efficient scavenger of solutes than is rain because the flakes have a high surface area. Thus, these have higher ionic concentrations. In particular, it is a very efficient scavenger for nitrate ions or nitric acid, so snowfalls can be expected to have more nitrate than has rainfall.

Dew has a very different composition from rainfall. The water that condenses upon leaves accumulates ions, such as potassium, calcium and magnesium, from the plant surface. Besides this, plants actively exude liquid from the leaf interior, in a process known as guttation, which is a strong source of chloride anions on the leaf surface.

5.5   Chemical reactions in droplets

Dissolution of soluble atmospheric trace gases in a small volume of liquid water considerably enhances their concentration. Once in droplets there are many new opportunities for reactions, although the low volume of the liquid phase means that only highly soluble gases will partition into the droplets. However, where gas phase reactions are slow, and the aqueous phase reactions comparatively fast, even a rather insoluble gas might be removed from the atmosphere effectively through aqueous reactions. This appears to take place in the removal of carbonyl chloride (i.e. phosgene) from the atmosphere. It is stable in the atmosphere, and is not very soluble. A low K_H of $0.1 \, mol \, l^{-1} \, atm^{-1}$ at $15 \, ^{\circ}C$ means that just a small fraction is present in atmospheric water. Yet it can be removed to raindrops because carbonyl halides hydrolyse rapidly:

$$COCl_2 + H_2O \rightarrow CO_2 + 2HCl \qquad (R5.13)$$

The oxidation of sulphur dioxide has been one of the most frequently studied reactions in aqueous atmospheric droplets. The gas is quite soluble when the subsequent hydration steps are considered and its oxidation product sulphuric acid is often responsible for acidifying rain. It has long been realised that the oxidation of aqueous sulphur dioxide by oxygen is very slow without catalysts. The only way in which this oxidation could be fast enough to be important in atmospheric droplets would be through the presence of catalysts such as iron or manganese. At acidities typical of atmospheric aerosols (about pH 5) sulphur dioxide will be present mainly as the bisulphite ion (HSO_3^-), so we might represent the oxidation by the equation

$$HSO_3^- + \tfrac{1}{2}O_2 \xrightarrow{Fe,Mn} SO_4^{2-} + H^+ \qquad (R5.14)$$

Manganese can be a very efficient catalyst at some pH values, but it is not as abundant as iron. The relative importance of these catalysts will depend on the origin of the droplet.

In remote areas the concentrations of these catalysts, although they are abundant from crustal sources may not be in a form that is soluble enough to promote oxidation. Here hydrogen peroxide and ozone are likely oxidants. Although they are present at low concentrations, there can be a sufficient amount of them to oxidise to sulphur dioxide in unpolluted areas.

$$HSO_3^- + H_2O_2 \rightarrow SO_4^{2-} + H_2O + H^+ \qquad (R5.15)$$

$$HSO_3^- + O_3 \rightarrow SO_4^{2-} + O_2 + H^+ \qquad (R5.16)$$

The product of these reactions is sulphuric acid. That it is a much stronger acid than sulphurous acid is the reason for the appearance of the proton on the right-hand side in each of the equations above.

The production of further acidity during oxidation is significant, because it means that, as oxidation proceeds, the droplet pH falls below the values expected from the presence of sulphurous acid alone: first, from production of the sulphuric acid, and second, from the dissolution of more sulphur dioxide to replace that which has been oxidised. The increase in acidity means that sulphur dioxide becomes less soluble as the reaction proceeds (i.e. the hydrolysis equilibrium lies more to the left-hand side – see the equations in section 5.2). In the case of the catalysed reaction (e.g. with iron as catalyst) the rate of oxidation also slows with increases in acidity. Therefore, the oxidation can rapidly come to a standstill. However, hydrogen peroxide is a very effective oxidising agent of sulphur dioxide in the atmosphere. This is because the rate at which it oxidises sulphur dioxide increases under acidic conditions, so that, as the reaction proceeds and more sulphuric acid is produced, the oxidation rate increases. The rates of these reactions as a function of pH are summarised in Fig. 5.3. If the reactions were to become very fast then the rate of

Fig. 5.3. Effectiveness of various aqueous sulphur dioxide oxidation mechanisms. Conditions are: $SO_{2(g)}$, 5 ppb; $H_2O_{2(g)}$, 1 ppb; $O_{3(g)}$, 50 ppb; $Fe(III)_{aq}$, 3×10^{-7} mol l^{-1}; and $Mn(II)_{aq}$, 3×10^{-8} mol l^{-1}. Data are from Seinfeld (1986).

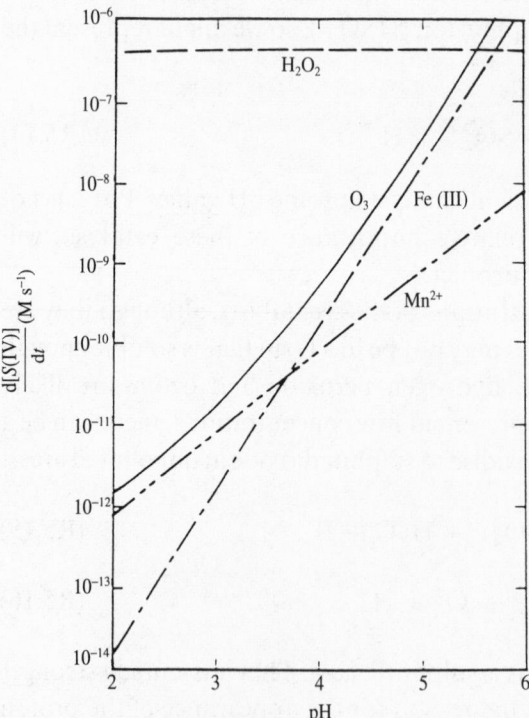

reaction could be limited by mass transfer processes. With oxidation by O_3 this would happen only when the pH rose above 6, while H_2O_2 reactions become mass-transfer-limited when the pH falls below 2.

The sulphur(IV) chemistry in droplets can be affected by other reactions. In particular, reactions with formaldehyde may be especially important:

$$H_2C(OH)_2 + HSO_3^- \rightarrow CH_2(OH)SO_3^- + H_2O \qquad (R5.17)$$

$$H_2C(OH)_2 + SO_3^- \rightarrow CH_2(OH)SO_3^- + OH^- \qquad (R5.18)$$

These reactions can enhance the effective value of the coefficient of Henry's Law of either formaldehyde (written in the equations as its hydrated form, the methylene glycol) or sulphur dioxide. The one present at the lowest concentration will be more effectively complexed by these processes. It is also possible for the metal ion catalysts of sulphur dioxide oxidation such as Fe(III) to react in solution and thus be rendered less efficient. This may be especially true of some organic acids that occur in solution and are likely to form complexes with transition metal ions.

Nitrogen oxides are also oxidised in droplets to form nitric acid, but the oxides have relatively low solubility. There is a possibility that some dissolution and reaction of two nitrogen oxides proceeds:

$$NO_2 + NO + H_2O \rightarrow HNO_2 + HNO_3 \qquad (R5.19)$$

The production of sulphuric or nitric acid results in an important subsequent reaction if the chloride ion is present in high concentration. Hydrogen chloride proves more volatile than other strong acids found in aerosol droplets, so it can be lost from the droplet according to the equation

$$H^+_{(aq)} + NaCl_{(asol)} \rightarrow HCl_{(g)} + Na^+_{(aq)} \qquad (R5.20)$$

Evidence for the occurrence of this reaction has been found by looking at the sodium-to-chloride ratio in maritime aerosols. Chloride concentrations are depleted when the sulphate concentrations are high. However, describing the reaction in a quantitative sense is complex, because it occurs in droplets that have nearly evaporated to dryness. These aerosols will have very high salt concentrations and the solution behaviour departs from ideality. This means that there are significant departures from thermodynamic predictions made using equilibrium constants obtained from low concentrations, without correcting with activity coefficients that are no longer close to unity.

It has been suggested that droplets in the atmosphere can also have an active radical chemistry. The hydroxyl radical barely survives for a second in the gas phase, so there is little chance of it dissolving in droplets. The HO_2 radical is relatively long-lived and soluble, so it can be scavenged from the gas phase by droplets:

$$HO_{2(g)} = HO_{2(aq)} \qquad (R5.21)$$

The constant of Henry's Law for this process is $9 \times 10^3 \, mol\,l^{-1}\,atm^{-1}$. Although this means that the radical is fairly soluble, dissolution can be further enhanced by hydrolysis reactions:

$$HO_2 \rightarrow H^+ + O_2^- \qquad (R5.22)$$

$$2H^+ + 2O_2^- \rightarrow H_2O_2 + O_2 \qquad (R5.23)$$

The acid dissociation (R5.22) has a pK_a of 4.69 so it will be important in alkaline or nearly neutral solutions. This is one source of aqueous HO_2, but these radicals are also often produced in solution.

Aqueous hydroperoxide radical can oxidise sulphur dioxide directly:

$$HO_2 + SO_3^{2-} \rightarrow SO_4^{2-} + OH \qquad (R5.24)$$

or via the production of hydrogen peroxide (i.e. reactions (R5.23) and (R5.15)).

Although the short lifetime of the OH radical means that it is not likely to be transferred to solution, there are possibilities by which it can be produced within droplets:

$$Fe(II)^{2+} + H_2O_2 = Fe(III)^{3+} + OH^- + OH \qquad (R5.25)$$

The aqueous chemistry of these radicals is not easy to study given their low concentrations. Nevertheless, the possibility of radical chemistry opens up the possibility of there being a whole range of organic reaction chemistry also. One piece of evidence that would support this novel chemistry is the presence of Cu(I) in droplets. Under oxidising conditions typical of solutions in equilibrium with the atmosphere copper is expected as Cu(II). Yet studies of droplets reveal copper to be in the lower valence state. The following reactions would be expected to give this:

$$Cu^{2+} + O_2^- \rightarrow Cu^+ + O_2 \qquad (R5.26)$$

$$Cu^{2+} + HO_2 \rightarrow Cu^+ + H^+ + O_2 \qquad (R5.27)$$

or perhaps even

$$Cu^{2+} + RCHO \rightarrow Cu^+ + RCO + H^+ \qquad (R5.28)$$

which gives rise to an organic radical in solution.

It is also possible that radicals can be photochemically produced within atmospheric droplets. One mechanism involving dissolved iron(III) is

$$Fe(OH)_n^{(3-n)+} \xrightarrow{h\nu} Fe(OH)_{n-1}^{(4-n)+} + OH + e_{(aq)}^- \qquad (R5.29)$$

offering another route for the formation of the aqueous OH radical and produces a short lived aquated electron. We have already noted that organic compounds are effective complexing agents for iron in droplets and can lower its effectiveness as a catalyst in S(IV) oxidation in solution. However, the oxalate, glyoxylate and pyruvate complexes of iron may enhance sulphur(IV) oxidation by encouraging a photochemically mediated production of hydrogen peroxide:

$$(Fe(III)-L)^{2+} \rightarrow Fe(II)^{2+} + L^{\cdot} \qquad (R5.30)$$

$$L^{\cdot} + O_2 \rightarrow {\cdot}O_2^- + L^+ \qquad (R5.31)$$

$$2O_2^- + 2H^+ \rightarrow H_2O_2 + O_2 \qquad (R5.32)$$

Here the organic ligands are symbolised by L and are being transformed to oxidised organic material. Hydroxyl radicals produced in droplets as suggested by reactions (e.g. (R5.24) or (R.5.25)) are likely to be involved in further reactions such as the production of formic acid from formaldehyde in the aqueous phase:

$$^{\cdot}HCHO + H_2O \rightarrow H_2C(OH)_2 \qquad (R5.33)$$

$$H_2C(OH)_2 + OH \rightarrow HC(OH) + H_2O \qquad (R5.34)$$

$$HC(OH)_2 + O_2 \rightarrow HO_2 + HCOOH \qquad (R5.35)$$

Reaction (R5.35) is an aqueous source of HO_2. This possibility was mentioned when we discussed the dissolution of HO_2, (R5.21)–(R5.23), and these aqueous sources can be important. At low temperatures, soluble precursors to HO_2 are more soluble, and solution phase production dominates. At higher temperatures both dissolution and production of the radical in solution are important.

The methylene glycol derived from formaldehyde (R5.33) can also react with hydrogen peroxide:

$$H_2C(OH)_2 + H_2O_2 \rightarrow HOCH_2O_2H + H_2O \qquad (R5.36)$$

Hydroxymethylhydroperoxide ($HOCH_2O_2H$) is very soluble ($K_H = 6.5 \times 10^5 \, mol\,l^{-1}\,atm^{-1}$ at $15\,°C$) and remains strongly partitioned into droplet phases. It is phytotoxic and has sometimes been considered as one of the compounds within polluted rainfall that could damage forests.

Precipitation is an effective method of removing trace substances from the atmosphere. However, in a global sense it is possible that precipitation could lower the potential of the atmosphere to oxidise trace species. This can come about because water droplets will lower the gas phase concentrations of soluble trace species such as formaldehyde, which is a source of HO_2 in the gas phase. Droplets can also reduce the gas phase concentrations of other species by less direct means. Ozone, for example, is an extremely important oxidising gas in the atmosphere. Although it is not especially soluble, it can react in droplets with the plentiful O_2^- radicals, thus being removed from solution:

$$O_3 + O_2^- + H_2O \rightarrow OH + 2O_2 + OH^- \tag{R5.37}$$

Thus, when clouds are present, the soluble species dissolve, but also relatively insoluble trace gases, such as O_3, are reduced. A shift in the concentration of ozone can have profound effects on atmospheric chemistry.

Clouds have yet another role to play in atmospheric chemistry. Convective motions associated with thunderstorms distribute trace gases throughout the vertical extent of the troposphere. This means that gases at ground level can be removed to higher altitude, where they will be less likely to be removed from the atmosphere by dry deposition, but probably more liable to oxidation by ozone.

5.6   Further reading

Chameides, W. L. and Davies, D. D. (1982) The free radical chemistry of cloud droplets and its impact upon the composition of rain, *Journal of Geophysical Research* **87**, 4863–77.

Graedel, T. E. and Goldberg, K. I. (1983) Kinetic studies of raindrop chemistry – I. Inorganic and organic processes, *Journal of Geophysical Research* **88**, 10 865–82.

Liss, P. S. and Slater, P. G. (1974) Flux of gases across the air–sea interface, *Nature* **247**, 181–4.

Mason, B. J. (1975) *Clouds, Rain and Rainmaking*, Cambridge University Press, Cambridge.

6

○ ○

Sources of pollution

6.1 Combustion processes

The most significant source of anthropogenic air pollution is the combustion of fuels. Fuels are energy sources of which the most familiar are carbon or hydrocarbon compounds. Elemental hydrogen has been proposed as a fuel, but is not widely used. At high temperature, fuels are rapidly oxidised by atmospheric oxygen, yielding a large amount of energy as heat, together with the products of combustion, which are usually stable gaseous oxides. Thermodynamic considerations would lead us to expect the hydrocarbons to be oxidised to carbon dioxide and water during combustion, i.e. if we burnt methane the reaction could be written

$$CH_4 + 2O_2 \rightarrow CO_2 + 2H_2O \tag{R6.1}$$

On a global scale, both carbon dioxide and water are released in enormous quantities in combustion processes (see Table 6.1). The water is not particularly significant because natural fluxes of water to the atmosphere are very much larger than any of those due to human activity. On a purely local level, the condensation of steam in plumes can be an aesthetic nuisance or create a hazard by reducing visibility. Carbon dioxide, on the other hand, is currently emitted in quantities that are large enough to perturb the concentrations found in the entire atmosphere, and thus enhances the greenhouse effect. In locations where large amounts of fuel are burnt, the carbon dioxide concentration of the air may be considerably enhanced, but there are no striking effects from the gas as a local pollutant. Viewed in this naive way it is hard to see why the effect of combustion processes on the composition of the atmosphere is of so much concern to society.

Carbon compounds

The air pollution problem arises because carbonaceous fuels yield many more combustion products than carbon dioxide and water. Take, for instance, the burning of wood. A substantial amount of the carbon is not converted through to carbon dioxide. Some is only partially oxidised to carbon monoxide or else remains as elemental or organic carbon in particulate materials emitted as smoke. Naturally, this is undesirable because it represents incompletely burnt and therefore wasted fuel. Moreover the visible smoke may cause a variety of pollution problems and carbon monoxide is poisonous. The organic carbon species emitted with the smoke are often a particularly complex mixture of compounds produced by reactions at elevated temperatures and some are toxic or carcinogenic. Of particular concern in wood-burning are the polycyclic aromatic hydrocarbons, or PAHs (see Fig 4.10(*b*) for structures), which are produced at temperatures a little above 600 °C, which is typical of wood-burning stoves. If the temperatures become high enough then PAHs will be destroyed, so one route to reducing the emission of PAHs would be to increase the temperature at which the wood is burnt. In an open fire the temperatures are usually quite high, and it is stoves that burn wood at a relatively low temperature that favour the production of large amounts of PAHs. They are also frequent products of some industries, such as coke plants, and from transport.

This section will centre on combustion in the gas phase. This is a reasonable way to look at combustion because, even with solid and liquid fuels, volatilisation of the fuel at high temperature means that much of the chemistry associated with combustion takes place in the gas phase. However, one thing needs to be remembered about solid fuels: traditionally

Table 6.1. *Global emissions of the products of combustion*

Fuel	CO_2^a (10^{12} kg a^{-1})	H_2O^b (10^{12} kg a^{-1})
Oil	2.25	0.35
Coal	2.0	0.1
Gas	0.75	0.13
Wood	0.5	0.1
Agricultural[c]	2.0	0.35
Natural sources	100	55 000

[a]Expressed as weight of carbon.
[b]Expressed as weight of hydrogen.
[c]This is not strictly fuel, but arises in grassland fires and slash-and-burn agriculture.

they represent the major contribution to the smoke or particulate material load of the urban air. In Europe and North America coal-burning industries and homes were major sources of smoke, but changes in fuel use have caused these sources to decline. Often tight controls on industrial emissions have added to the reduction. In the UK, for example, by 1980, although the domestic emission of smoke was $0.3 \, \text{Tg} \, \text{a}^{-1}$, industrial emissions were even lower at less than $0.05 \, \text{Tg} \, \text{a}^{-1}$ (see Fig. 6.1). Some of the reduction over the last fifty years has been achieved through smoke abatement technology, but transition to cleaner fuels, such gas or electricity (produced with smoke removed by electrostatic precipitation), has been even more important. The domestic sector is still important in the 1980s, but it is road transport that produces almost half the black smoke emission. Thus, much of the soiling potential of urban air is produced by an increasing diesel fleet.

Of the gaseous carbon compounds, it is carbon monoxide that is the major pollutant, with large amounts being produced by automobiles. The production of carbon monoxide can be described in terms of the gas phase equilibrium when there is insufficient oxygen to burn a fuel:

$$CO + \tfrac{1}{2}O_2 = CO_2 \hspace{3cm} \text{(R6.2)}$$

$$K = p(CO_2)/[p(CO)p(O_2)^{0.5}] \hspace{3cm} \text{(6.1)}$$

Fig. 6.1. Black smoke by sector in the UK, 1970–1990 (source HMSO).

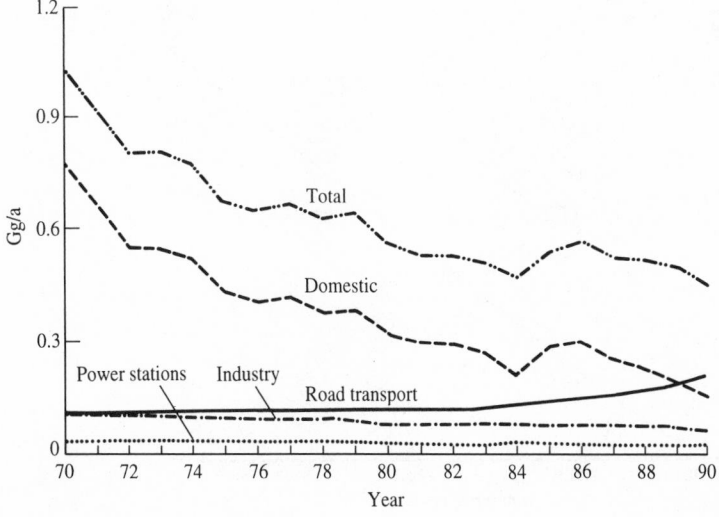

The values of this equilibrium constant as a function of temperature are shown in Fig. 6.2. High temperatures favour the formation of carbon monoxide and equilibrium considerations suggest that increasing the oxygen-to-carbon ratio will favour complete oxidation. This is found, but the internal combustion engine of the automobile generally runs on a fuel-rich mixture, i.e. with less oxygen than is needed to oxidise all the fuel to carbon dioxide and water, which favours carbon monoxide production. Thus automobiles represent an important source of carbon monoxide, although engineers have designed 'lean burn' engines with increased air-to-fuel ratios that help lower carbon monoxide emissions.

Strangely enough the carbon monoxide concentration found in exhaust gases are not at the value that one would expect from the equilibrium conditions. Temperatures during combustion are initially about 2500 K at a pressure of 70 atm. Even an understanding of the kinetic factors does not entirely explain the fact that the carbon monoxide concentrations found in exhaust emissions are higher than predicted by theory. One possibility is that the reactions, at least within the internal combustion engine, occur in a poorly mixed medium.

The discussion of carbon monoxide has centred on the motor vehicle. This seems reasonable because of the fuel-rich mixtures used in automotive engines. Although they do not always consume the most fuel, they can easily be responsible for the bulk of carbon monoxide found in urban air (Fig. 6.3).

Fig. 6.2. Values of the equilibrium constant for the reaction $CO + 0.5O_2 \rightleftharpoons CO_2$ as a function of temperature.

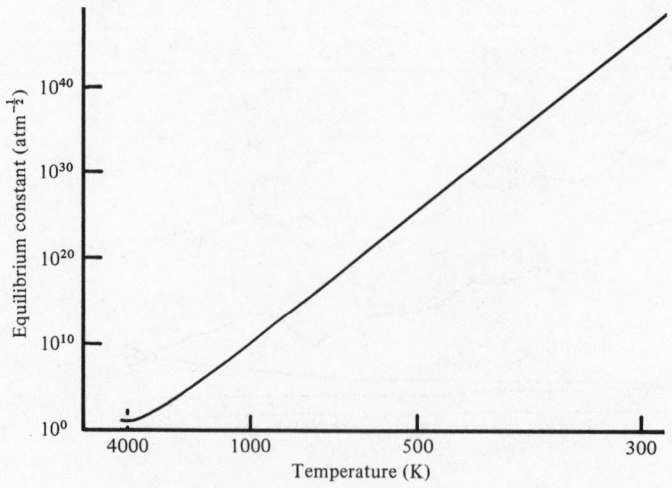

The unburnt hydrocarbons found in the exhaust gas from automobiles represent carbon even less oxidised than carbon monoxide. Theoretical considerations suggest that hydrocarbon molecules should be destroyed in combustion because the hypothetical equilibrium

$$C_8H_{18} + 12.5O_2 \rightleftharpoons 8CO_2 + 9H_2O \qquad\qquad (R6.3)$$

lies very much to the right-hand side. Besides this, kinetic studies show that the oxidation reactions of hydrocarbons are fast under the conditions found within the internal combustion engine. There has to be another reason for the presence of hydrocarbons in exhaust gases. An important clue to this comes from the fact that the hydrocarbons found in exhausts are not present in the same ratios as in the original fuel. The fuel hydrocarbons for a normal automotive engine consist of various paraffins containing between four and nine carbon atoms per molecule. However, many hydrocarbon compounds found in exhaust gases are much smaller molecules, typically methane, ethene (ethylene) and ethyne (acetylene). These are often all at concentrations greater than 100 ppm in automobile exhausts. Hydrocarbons of low molecular weight are probably formed in the much cooler layers of gas that exist against the wall of the engine cylinder during combustion. Pyrolysis reactions tend to produce unsaturated compounds, so ethane is likely to be found at concentrations perhaps ten times lower than that of ethene in exhaust gases. Acetylene is emitted from automobiles in large quantities. It is reasonably stable in the atmosphere and hence is an excellent tracer of air masses that have been polluted by automobiles. If temperatures in some region of the internal combustion engine are low enough then some organic compounds escape complete

Fig. 6.3. Carbon monoxide emissions by sector in the UK, 1990, with the total being 6.7 Tg (source HMSO).

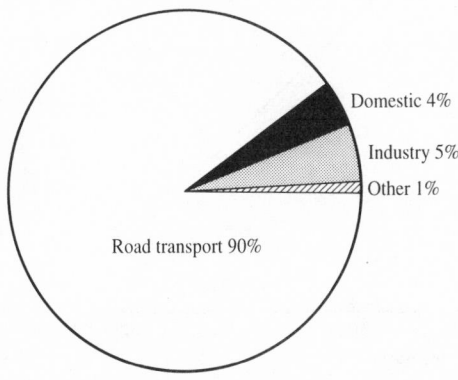

Domestic 4%

Industry 5%

Other 1%

Road transport 90%

destruction. In particular, relatively high concentrations of pentanes, benzene, toluene and the xylenes are found in exhaust gases.

Nitrogen compounds
The description given above leaves us with a range of carbon compounds and water as the principal emission products of combustion of a typical fuel. However, the discussion considered combustion in the presence of oxygen, whereas it takes place in air that is predominantly nitrogen. The high temperatures in the combustion process encourage the formation of oxygen atoms that can enter chain reactions:

$$O + N_2 \rightarrow NO + N \qquad\qquad (R6.4)$$

$$N + O_2 \rightarrow NO + O \qquad\qquad (R6.5)$$

These can be summed to give the overall process

$$N_2 + O_2 \rightarrow 2NO \qquad\qquad (R6.6)$$

This equilibrium lies very much to the left-hand side at ambient temperatures (see Fig. 6.4), but at 2500 K the equilibrium constant is

Fig. 6.4. Values of the equilibrium constant for the reaction $N_2 + O_2 \rightleftharpoons 2NO$ as a function of temperature.

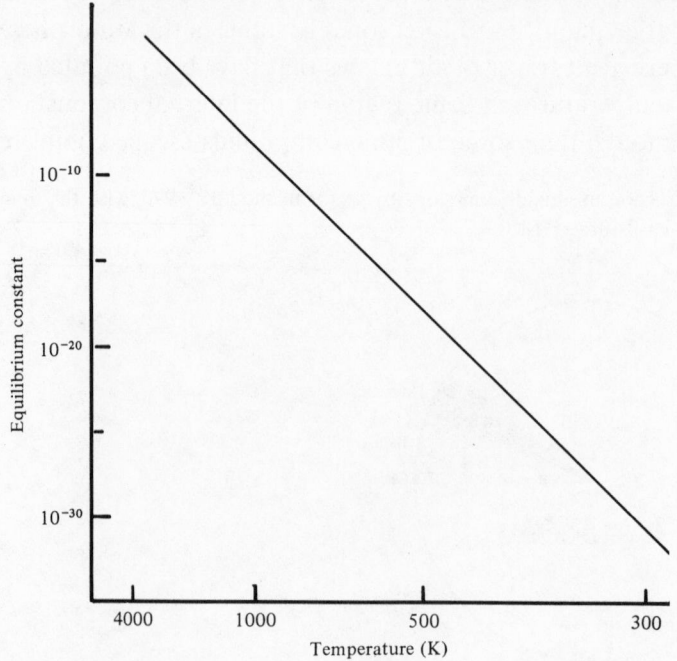

3.5×10^{-3}. Thus, high temperatures encourage the formation of nitric oxide. In an internal combustion engine the rate of the chain reactions is too slow to allow equilibrium concentrations of nitric oxide to be reached. However, there are more rapid routes to produce nitric oxide during the combustion of fuels. In the combustion of hydrocarbons, the presence of the CH radical can lead to rapid formation of HCN:

$$CH + N_2 \rightarrow HCN + N \qquad\qquad (R6.7)$$

which is then converted to CN and the atomic nitrogen is oxidised to NO. In addition to this, when the fuel actually contains some organic nitrogen, it can be converted to ammonia or hydrogen cyanide, which oxidises to NO.

As the exhaust gas cools, the equilibrium shifts and low concentrations of nitric oxide would be expected, but the rate of equilibration becomes extremely slow as the gas cools, so the equilibrium is 'frozen' at a high-temperature value. The fact that nitric oxide originates from high-temperature reactions in air means that it occurs in every high temperature combustion process in air, so even the cleanest fuel is a source of pollution from these compounds. Running an engine on a hydrogen–air mixture is very non-polluting (i.e. no carbon-containing pollutants, only water). However, it still gives rise to nitrogen oxides, although with careful control these can be an order of magnitude lower in amount than those from gasoline. This means that the claim that the use of hydrogen as a fuel is pollution-free is not particularly exaggerated.

Power stations and industry have traditionally made the largest contribution to total nitrogen oxide emissions. In recent years the

Fig. 6.5. Nitrogen oxides emissions, expressed as nitrogen dioxide equivalent, by sector in the UK, 1990. The total emissions amounted to 2.7 Tg (source HMSO).

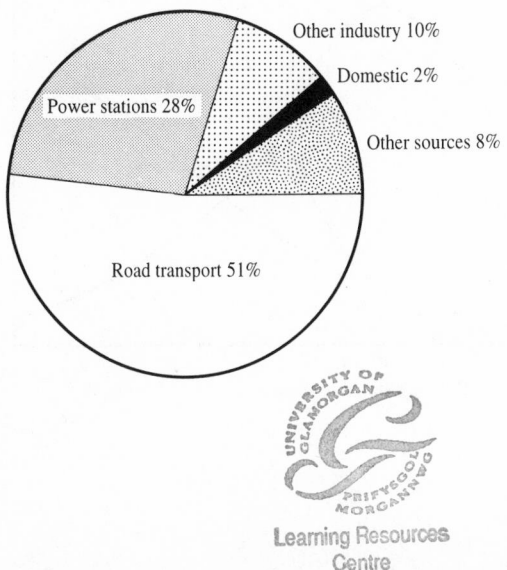

automobile has come to represent the dominant source of nitrogen oxides, in much the same way as it does for carbon monoxide (see Fig. 6.5). However, there is considerable interest in controlling nitrogen oxide emissions from automobiles because in cities they represent a concentrated urban source, which bears much of the blame for photochemical smog. One difficulty in controlling nitrogen oxide emissions is that mixtures with low concentrations of oxygen (rich mixtures), which reduce the emissions of nitrogen oxides, enhance the emission of carbon monoxide and hydrocarbons (Fig. 6.6). Changes in engine design, the addition of catalytic converters or after burners and careful tuning considerably reduce the major pollutants in exhaust gas. The difficulty with these approaches is that they often work well on a test bed with warm engines under suitable operation conditions, but are not always so effective in poorly maintained vehicles operating under stop–start city driving.

6.2   Incineration and impurities in fuels

Nitrogen present as an impurity in fuels can give rise to pollutant nitric oxide in exhaust gases. In a similar way, other trace impurities can give

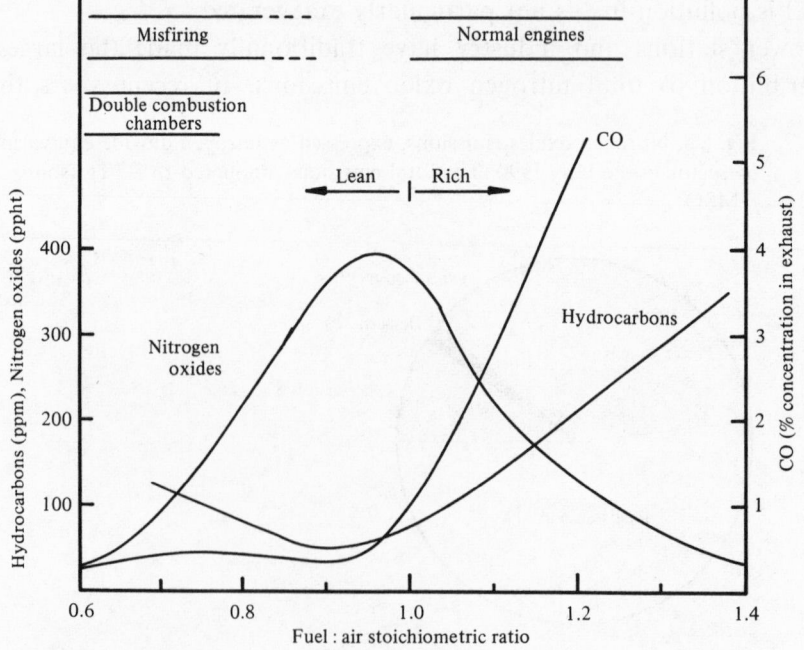

Fig. 6.6. Concenrations of pollutant gases in automobile exhaust as a function of the fuel : air ratio. ppht = parts per hundred thousand (from Perkins (1974) and Campbell (1977)).

rise to a variety of pollutant gases in emissions. The presence of chlorine and sulphur in fuels results in the emission of gaseous chlorine and sulphur compounds.

Sulphur compounds

Sulphur has traditionally been seen as the most important of the trace impurities in fuel because of its particularly high concentrations in some coals (as high as 6%). Combustion readily causes its oxidation to sulphur dioxide that has been such an archetypical urban air pollutant.

Sulphur is present in many fuels over a wide range of concentrations. In coal it is present as iron pyrites and sulphur-containing organic compounds in roughly equal amounts. Crude oils contain anything from a few tenths of a percent to about 2% sulphur. During refining, much of the sulphur often remains with the heavier petroleum fractions, so that the sulphur content of residual fuel oils can be a few times higher than that of the crude oils. If demand for a fuel such as diesel is particularly high then there may be a need to use a lower grade feedstock and its sulphur concentration and general quality could decrease. In recent years petroleum companies have tried to maintain the quality of diesel fuel, while legislation has defined upper limits to its sulphur content within many countries.

Natural gases usually contain relatively low concentrations of sulphur. Small amounts of sulphur are added, usually tertiary-butylmercaptan (2-methyl-2-propanethiol), as an odourant. This substance can be smelt at concentrations as low as 0.5 ppb and enables leaks to be detected easily in the domestic situation. Coal gas, which is largely a mixture of hydrogen, carbon monoxide and hydrocarbons, is produced from distillation of coal. It is also low in sulphur because the hydrogen sulphide present is traditionally removed by passage over ferric oxide and, in modern times, by using a variety of amines. Desulphurisation of coal gas was required by early Gas Acts passed in the UK during the last century. Thus the conversion of a fuel, such as coal, to gas can be an effective method of lowering emissions of sulphur dioxide. Some newer methods of obtaining gas from coal look very promising as they give greater yields than conventional methods, e.g.

$$C + H_2O \rightarrow CO + H_2 \qquad \text{(R6.8)}$$

$$CO + H_2O \rightarrow CO_2 + H_2 \qquad \text{(R6.9)}$$

The coal is represented by C in the first reaction. The first of these two reactions may be familiar as it represents the production of 'water gas' or 'producer gas', which has traditionally been used as a cheap fuel for firing large furnaces. However, the two reactions can be carefully controlled to achieve a ratio of hydrogen to carbon monoxide of $3:1$ and may be followed by

$$3H_2 + CO \rightarrow CH_4 + H_2O \tag{R6.10}$$

This may prove an effective way of producing a non-polluting fuel from abundant high-sulphur coals. The glut of low-grade fossil fuels in the 1990s has spurred an unprecedented development of integrated gasification combined cycle plants for generating electricity.

Chlorine compounds

Even a fuel as innocuous as wood can release about $2\,mg$ of methyl chloride for every gram burnt. Coals burnt in power generation can be highly variable in their chlorine content, so that some will give substantial amounts of hydrochloric acid. Large quantities of hydrogen chloride can also be generated during incineration of plastics such as polyvinylchloride. This means that refuse burning on building sites and in municipal incinerators represent important sources of hydrochloric acid. Their emissions need careful control. The corrosiveness of the exhaust gas from such plants often creates a challenge for designers who have to ensure that the internal surfaces have a reasonable lifetime in such an environment.

Chlorine proves an even more worrisome element during combustion because of the possibility that reactions within furnaces can lead to production of carcinogenic chlorinated compounds such as the poly-chlorinated dibenzodioxins and dibenzofurans (PCDDs and PCDFs). These include the notorious 2,3,7,8-tetrachlorodibenzo-*p*-dioxin (often known as TCDD or even more simply *dioxin*). Dioxin has received wide public attention after it was found as a contaminant in the herbicide 'agent orange' used during the Viet Nam war. Further interest arose when it was formed and released during an industrial accident at Seveso in Italy in 1976. These chemical sources of dioxin are now highly regulated, so there has been a shift of interest towards the less obvious sources.

Some PCDDs and PCDFs can be released from incinerators because traces may have been present in the material being burnt. However, experiments in which material that contains chloro-organics, but no PCDDs and PCDFs have shown that they can be readily synthesised during combustion of fairly innocuous substances. It is possible that elemental carbon can react with incinerator ash to produce chlorinated aromatic compounds. Chlorophenols that can be formed under conditions in which combustion is incomplete can act as precursors for the formation of dioxin. A simple representation of the process can be written as

(R6.11)

Copper(II) chloride, which is an excellent oxychlorination catalyst, is one component of ash that may aid the formation of PCDDs and PCDFs. The formation of furans is favoured by higher concentrations of copper in ash, so this could explain the tendency of incinerator emissions to contain both PCDDs and PCDFs. Where fuels contain chlorine and particle surfaces for catalysis, some dioxin production is to be expected. Domestic heating with fuel oil and automobiles that use leaded fuel (in which chlorine is an additive) are likely urban sources of dioxin. Agricultural burning can also produce dioxins because crop residues also contain chlorine, some perhaps as pesticides. However, processes such as straw burning, which is now banned in the European Union, produced very small amounts of PCDDs and PCDFs.

Metals

Metals are also found as impurities in fuels. The concentrations of metallic trace elements in fuels are highly variable. For instance, in some crude oils vanadium is found in concentrations as low as 0.1 ppm while in others it may be at concentrations as high as 1400 ppm. In Fig. 6.7 are shown typical concentrations of both metallic and non-metallic trace elements in coal and oil as a function of their crustal abundance. The elements lying above the diagonal lines are those that are enriched in the fuels relative to the crust. Trace element concentrations are much lower in oil than in coal. The striking exception to this generalisation is vanadium, in which case considerable enrichment has taken place. Nickel and mercury are also enriched in oil. Metals in oils can be found as metalloporphyrin chelates, complexes of tetradentate ligands, organo-metallic compounds and salts of carboxylic acids. Vanadyl porphyrins are exceptionally stable which accounts for the large concentrations of vanadium found within oils. Thus, vanadium is an excellent tracer for emissions from oil combustion. It is present in oil fly ashes as a range of oxides (V_2O_5, V_2O_4 and V_2O_3) of different valency and as mixed oxides with sodium and more notably with nickel as $2NiO \cdot V_2O_5$. The presence of a mixed oxide with nickel seems a natural outcome of an enrichment often as pronounced as that for vanadium.

Fig. 6.7. Elemental abundances in coal and oil compared with abundances in the Earth's crust. Diagonal lines indicate the position that would be occupied by elements showning no enrichment relative to iron.

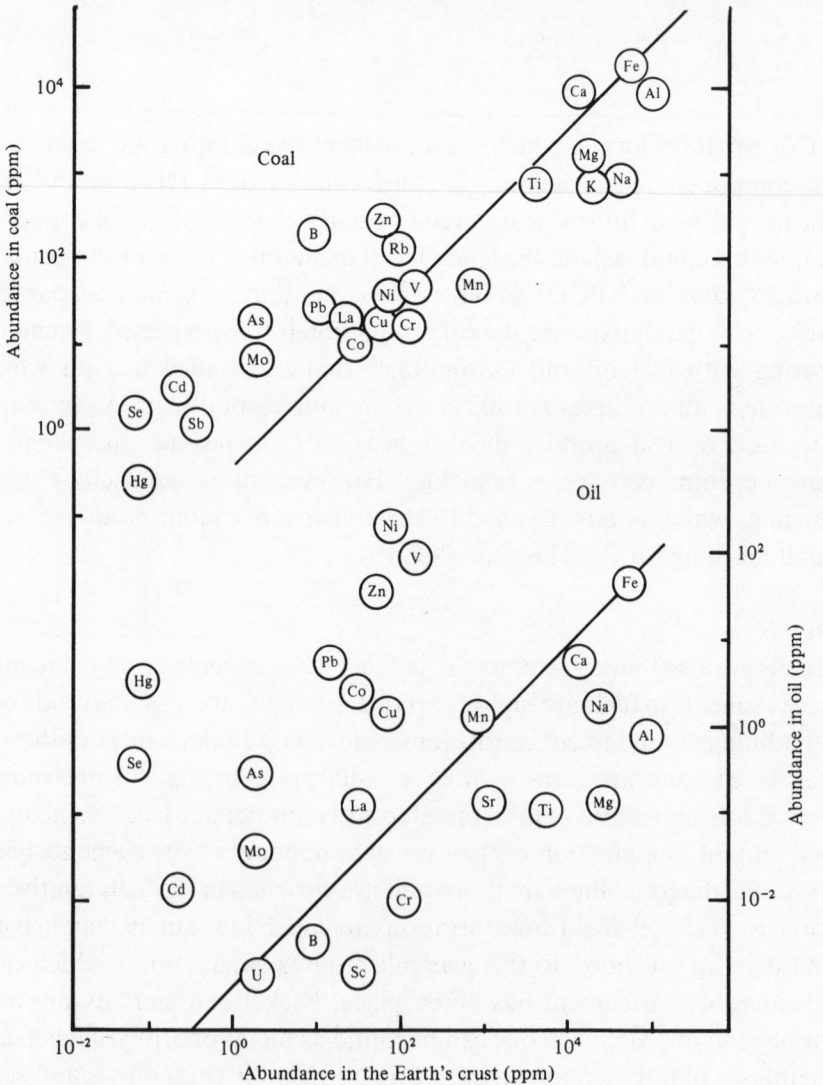

The presence of toxic trace elements in coal, coal ash and smoke has been of interest for many hundreds of years. In the 1600s scientists wondered whether the high death rate in London might not be due to toxic trace elements from the combustion process. More than a century ago it was known that the coal-smoke-laden air contained measurable concentrations of arsenic. The possible dangers from these elements have recently been placed in sharper perspective. Lead, thallium, antimony, arsenic and mercury have been discovered at enhanced concentrations on the surface of coal ashes. The enhancement is highest on the smaller particles. Such particles remain in the atmosphere for longer and are more likely to penetrate deep into the lung in inhaled air.

Municipal incineration also introduced many metals into the combustion process. Some of these, such as copper, were discussed before as potential catalysts for the formation of chlorinated organic compounds. Metals with lower boiling points, such as cadmium, manganese, antimony, lead and mercury, are likely to be found in the vapour phase in furnaces. Mercury can remain in the vapour phase at ambient temperatures, but the others are likely to transfer to the surface of fine particles as they cool.

The toxic metals that are most readily enhanced usually have the higher volatility (see Section 4.3). This is why it is so generally assumed that they are distilled onto fine particles of ash during the thermal changes that occur as particles proceed from the interior of the furnace into the ambient air. Nickel and chromium are also found at enhanced concentrations on the surface of fly ashes. However, these elements are not particularly volatile, so distillation does not seem a very plausible mechanism. It is possible that the enrichment arises from transfer of these elements to particle surfaces as volatile sulphides or possibly carbonyls (although carbonyls such as $Ni(CO)_4$ and $Cr(CO)_6$ have limited stability at high temperature).

These pictures of the enhancement of metals through vapour transport suggest increased concentrations on the surface of the particles, with little enhancement in the core. Small particles should show a higher enhancement because they have a higher surface-area-to-volume ratio and this has been observed sometimes. The existence of an enhancement mechanism that places high concentrations of volatile and toxic metals on the surface of small and easily inhaled particles may mean that the metals are readily available and could present a health hazard.

Not all the trace metal components of fuels are of natural origin. Some may be the result of contamination during processing, while others are deliberately added. Lead compounds are added to petrol, for example.

The addition of tetra-ethyl lead as an anti-knock agent that allows the fuel to burn correctly within the engine cylinder is well known. Automotive fuels have contained lead at concentrations that have been in excess of 0.5 g l^{-1} and such large values are still to be found in the fuels of many developing countries. Concern about the dispersion of enormous amounts of this toxic metal in the environment has resulted in a reduction in the permitted levels of this additive. Ethylene dibromide and ethylene dichloride are added to leaded petrols as scavenging agents, which can result in the emission of lead bromochloride from motor vehicles. An increasing fraction of the vehicle fleet now operates on petrol without added lead. This has meant a continued decline in the total emission of lead (see Fig. 6.8). In Canada, methylcyclopentadienyl manganese tricarbonyl has replaced lead as an anti-knock agent. This leads to the formation of Mn_3O_4 particles in the atmosphere, but they probably do not cause undesirable increases in atmospheric manganese concentrations.

6.3   High-temperature sources

High-temperature processes such as smelting are a large source of air pollutants. The high temperatures are frequently achieved by combustion of fuels, so pollutants are obviously generated in the ways discussed in the previous section. However, there are many other high-temperature chemical reactions that form pollutant gases and particles. Take the production of cement from clay and limestone, which may be written

$$CaAl_2Si_2O_8 + 6CaCO_3 \rightarrow 2Ca_2SiO_4 + Ca_3(AlO_3)_2 + 6CO_2$$

$$(R6.12)$$

Fig. 6.8. Emissions of lead in the UK (source HMSO).

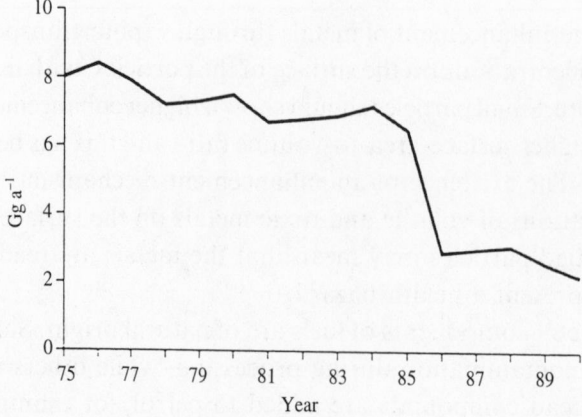

The manufacture of cement leads to the production of massive amounts of carbon dioxide. Global production from cement is in the region of a billion tonnes per year. This represents a small contribution to the flux of carbon dioxide to the atmosphere from human activities.

Metal ores often occur as oxides or sulphides and must be reduced to the elemental state during smelting and refining. For instance, iron oxide can be reduced with carbon or carbon monoxide:

$$3Fe_2O_3 + CO \rightarrow 2Fe_3O_4 + CO_2 \tag{R6.13}$$

$$Fe_3O_4 + CO \rightarrow 3FeO + CO_2 \tag{R6.14}$$

$$FeO + C \rightarrow Fe + CO \tag{R6.15}$$

$$FeO + CO \rightarrow Fe + CO_2 \tag{R6.16}$$

The annual production of iron is about a billion tonnes per year and also represents a minor global source of carbon dioxide.

More serious are the large quantities of sulphur dioxide produced from the oxidation of sulphide ores in the production of metals such as nickel. Sulphide ores can be roasted to give the oxide that is then reduced. The production of sulphur dioxide can be understood from the equation

$$Ni_2S_3 + 4O_2 \rightarrow 2NiO + 3SO_2 \tag{R6.17}$$

In smelting operations the high temperatures allow an effective vapour phase transfer of some trace substances to particulate material. This may be the metal being extracted or other components of the ore. Ores can contain high concentrations of volatile and toxic substances. For instance, high concentrations of arsenic are often associated with nickel ores. As early as the sixteenth century Agricola, the German mineralogist, pointed out the health hazards associated with metal extraction. In seventeenth century England much concern was expressed over the release of arsenic and other compounds into the air during smelting operations. They were considered the cause of regional diseases that were prevalent in regions near smelters.

Aluminium smelting and brick-making are important sources of fluoride pollution. In aluminium smelting the fluoride comes from the cryolite ($AlF_3 \cdot 3NaF$) used as a solvent for alumina. Production of bricks required that clays be baked to drive off water. Fluorine can substitute for the OH group in clays, so hydrogen fluoride can be driven off rather than water at high temperature. The clays used in brick-making are often very complex mixtures. However, the production of hydrogen fluoride may be understood from the equation below, which represents the high-temperature reaction between a muscovite and silica that yields microcline, sillimanite and hydrogen fluoride:

$$K_2Al_4(Si_6Al_2O_{20})(OH)_2F_2 + 2SiO_2$$
$$\rightarrow 2KAlSi_3O_8 + 2Al_2SiO_5 + 2HF \qquad (R6.18)$$

Some mercaptans can also be produced during the manufacture of bricks and these give rise to the rubbery odour that characterises areas of brick-production. The smell of mercaptans is one of many odours that can arise in food processing.

6.4   Low-temperature sources

It is also possible to have low-temperature sources of air pollutants. Increasingly it is the evaporation of volatile organic compounds (VOCs) that is of most concern. In the UK the total emissions are estimated at just on $3\,Tg\,a^{-1}$. The breakdown of sources of VOCs (neglecting methane) is shown in Fig. 6.9(a). Obviously, solvent use and the petrochemical industries are responsible for a large contributions to the VOC budget. Transport generates substantial emissions through fuel use and its evaporation. The emissions are large, but there are many moves underway to lower these emissions. Catalytic converters that oxidise the organic components of exhaust gases and methods to help lessen evaporative losses from engines and fuel tanks are increasingly common. Fugitive emission of vapours in the chemical industry can be lessened by choice of better seals and valves. Often the emissions represent a substantial loss of material, so expenditure on better seals can repay itself. There is a broadening use of aqueous solvents that has helped reduce VOC emission in some industrial sectors.

Of the VOCs to be found in urban atmospheres, it is notable that the aromatic solvents benzene and toluene make up about 10% (toluene being the larger by a factor of three). The high reactivity of toluene and the carcinogenicity of benzene make the size of these emissions a matter for concern. The lower alkanes ($<C_5$, but excluding methane) make up another 20% of the emissions, although they are less reactive and not so harmful to human health.

Gas leakage is now a minor source of VOCs in the UK at 1% of the total. However, it is a more significant source of methane (Fig. 6.9(b)). The petroleum industry and mining in the UK are important sources, but it is agricultural animals that release a surprisingly large amount (it may be as much as $1.2\,Tg\,a^{-1}$). A further large source is the production of methane in landfill sites. There has been a rise in the incidence of methane accumulation in homes that are near to waste and landfill sites and there have also been some explosions.

A few substituted alkanes are produced in very large quantities. Global production of carbon tetrachloride has sometimes been as high as $0.1\,Tg\,a^{-1}$. The production of chloroform is estimated at as much as $1\,Tg\,a^{-1}$. The production of freons as refrigerants and aerosol propellants is lower (about 0.35 and $0.4\,Tg\,a^{-1}$ for Freon-11 (CCl_3F) and Freon-12 (CCl_2F_2) respectively in the 1970s). However, their high stability has led to considerable concern, particularly regarding the effect that they might have on stratospheric chemistry (see Section 8.1). Together with this, bromoform and halons used in some fire extinguishers (e.g. $CBrClF_2$, $CBrF_3$) are now additional sources of halogenated compound on a global scale.

Large amounts of bromodichloromethane and dibromochloromethane are released by reactions that occur in solution after the chlorination of drinking water supplies. Neither of these volatile substances is used in large quantities industrially. They de-gas from waters over time, which is probably the major reason for their high concentrations, sometimes as much as 100 ppb, in populated areas.

The food industry is associated with many odours. The smell of roasting coffee or chocolate is pleasant enough in small doses, but there are frequent complaints from residents of neighbourhoods downwind from factories making coffee or chocolate. The odours from brewing and

Fig 6.9. Non-methane VOC (*a*) and methane (*b*) emissions by sector in the UK, 1990 (source HMSO).

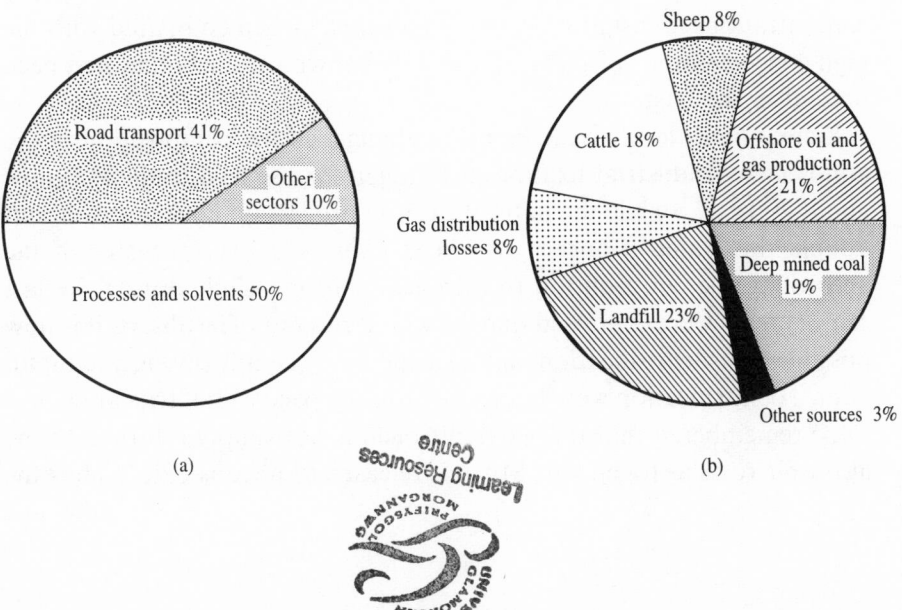

(a)   (b)

meat processing are particularly offensive. Reduced sulphur gases are often responsible for the odours from putrefaction and are also linked to complaints about gases from waste sites, where they are more easily perceived than the explosive methane emissions.

Another large source of atmospheric pollution is agriculture. Here the emissions are wide-ranging. Some can be seen as the product of combustion. 'Slash and burn' agriculture makes a large contribution to the global particulate carbon emission. This amounts to some teragrams per year, although it is not always easy to decide whether the emissions are the result of agricultural activities, accident or natural fires. Changes in European law now limit the extent of stubble burning which used to lead to high atmospheric particulate concentrations in the summer months. Any failure in care for soils can result in erosion, loss of vegetation cover and enhanced release of dust. Unsealed road surfaces can also be the source of large amounts of dust.

Agricultural dressings, pesticides and herbicides now form an important part of the anthropogenic contribution to the atmosphere, particularly as organohalogen and organometallic compounds. The best known of these compounds is DDT, which achieved a peak production of 8×10^{10} g in the early 1960s. It was applied to soils at quite high loadings and may be found there at application densities of many grams per square metre. The mean lifetime of DDT in the soil is about five years. It is gradually removed by evaporation, harvesting of the crops, leaching by drainage water and through degradation. The pesticide is reasonably volatile and the atmospheric concentration is thought to have reached a maximum of $80\,\mathrm{ng\,m^{-3}}$ in the mid-1960s. However, now it is the more modern pesticides that appear ever more widely dispersed over the globe, but their concentrations areusually low in human tissue when compared with the high concentrations of DDT and its breakdown products that had been present in the past.

Increasing of fertility can result in changes in the composition of the atmosphere. Industrial fixation of nitrogen for use as a fertiliser is now about $70\,\mathrm{Tg\,a^{-1}}$ while the natural fixation process on land is only about double this value. Concern has been expressed over the size of the anthropogenic contribution to nitrogen fixation. While during the last century fears were expressed that we would run out of fertilisers, it is now possible that denitrification cannot cope with the anthropogenic input. Even if denitrification were to increase to keep pace with nitrification, it is to be remembered that this extra nitrogen is not simply returned to the atmosphere as nitrogen gas. Much is released as nitrous oxide. Thus the increasing intensity of agriculture is probably enhancing the nitrous oxide emissions and changing the chemistry of the troposphere and stratosphere.

6.5   Photochemical pollution

As discussed in earlier chapters, reactions that take place in the atmosphere represent a significant source of atmospheric trace substances. The same is true of pollutants. The trace gas concentrations in polluted atmospheres can be much higher than those found in the unpolluted atmosphere, so the rates of reactions, particularly second-order ones, can be considerably enhanced. Many compounds have been identified in polluted urban air, so the chemistry is potentially extremely complex. Much research has been devoted to studying the situation prevailing in the smog of the Los Angeles basin. The processes occurring in the air of Los Angeles model many similar situations in which there is low humidity, plenty of sunshine, a large amount of vehicular traffic and moderate to low wind speeds. Photochemical pollution is now more common than was originally thought. Many pollutants and reactions first characterised in the air of the West Coast of the USA have now been detected in polluted locations all over the world, despite remarkable differences in fuel usage and weather. Photochemical smog occurs so widely that it is important to discuss it in some detail.

Nitrogen oxide pseudo-equilibrium
At the root of the photochemical smog problem are the nitrogen oxides, NO and NO_2, which are sometimes written in sum as NO_x. Nitric oxide, as noted earlier, arises from the oxidation of atmospheric nitrogen in combustion. It is liable to further oxidation.

$$2NO + O_2 \rightarrow 2NO_2 \qquad R = k_1[NO]^2[O_2] \qquad \text{(R6.19)}$$

$$NO + O_3 \rightarrow NO_2 + O_2 \qquad R = k_2[NO][O_3] \qquad \text{(R6.20)}$$

The reaction of nitric oxide with oxygen at the concentrations found even in the air of Los Angeles is slow. The main way in which nitrogen dioxide is produced is by oxidation with ozone. Taking typical early morning values for these compounds in polluted air (for O_3, 40 ppb and for NO, 80 ppb) and the rate constants k_1 and k_2 of 1.93×10^{-38} cm^6 s^{-1} and 1.8×10^{-14} cm^3 s^{-1}, we get rates of 4.3×10^5 cm^{-3} s^{-1} and 3.9×10^{10} cm^{-3} s^{-1}, confirming the far greater importance of oxidation by ozone. The nitrogen dioxide produced in this way can photodissociate back to nitric oxide, so it is possible to write a sequence of reactions to describe its destruction and regeneration:

$$NO_2 + hv \text{ (less than } 430\,nm) \rightarrow O(^3P) + NO \tag{R6.21}$$

$$O(^3P) + O_2 + M \rightarrow O_3 + M \tag{R6.22}$$

$$O_3 + NO \rightarrow O_2 + NO_2 \tag{R6.23}$$

Consider a volume of air in steady state, where the production of NO_2 is equal to its rate of destruction. Assuming (R6.19) to be unimportant, this may be written as

$$k_2[NO][O_3] = J[NO_2] \tag{6.2}$$

where J is the effective first-order constant for photodissociation. This equation can be rearranged to give

$$J/k_2 = [NO][O_3]/[NO_2] \tag{6.3}$$

where the term on the right-hand side can be imagined as a pseudo-equilibrium constant relating the partial pressures of the two nitrogen oxides and ozone. Of course, the value of J varies through the day as the sunlight changes in intensity, but measurements have shown that, overall, the equality implied by this equation holds in the polluted atmosphere. Let us imagine what would happen as the radiation intensity changes throughout the first half the day. This means that J will increase and increasing amounts of O_3 and NO are to be expected. As these are both produced by the destruction of NO_2, the amount of ozone should approximately equal the amount of nitric oxide.

If we look at some measurements from polluted atmospheres then neither of these predictions is borne out. It is true that the concentration of NO does rise in the early morning, but that of ozone rises much more slowly. In even greater disagreement with our prediction is the fact that the NO levels fall by midday. Of course, one argument that one might make is that the rises and falls in pollutant concentration that we are seeing are merely functions of the pattern of generation and dispersion in the atmosphere. It is true that, in the Los Angeles basin, there is a peak in production of pollutants in the morning while many people are driving to work. However, results very similar to those shown in Fig. 6.10 are found for closed experimental smog chambers, in which the generation and dispersion of pollutants are under control.

The role of organic molecules
To understand the situation, we must consider the other components of

the polluted atmosphere. Under constant illumination, the rise in O_3 concentration shows a decreasing NO/NO_2 ratio in the pseudo-equilibrium. For this to happen there needs to be another source of oxidant, because the reaction sequence illustrated by reactions (R6.21)–(R6.23) does not result in any overall production of ozone. The production of ozone, occurs, as do many processes in the unpolluted atmosphere (Chapter 3), through reactions in which the OH radical plays a key role. The hydroxyl radical attacks a variety of pollutants in the urban air:

$$OH + CH_4 \rightarrow H_2O + CH_3 \tag{R6.24}$$

$$OH + CH_3CHO \rightarrow H_2O + CH_3CO \tag{R6.25}$$

$$OH + CO \rightarrow H + CO_2 \tag{R6.26}$$

As an initial simplification, we can regard methane as a typical alkane. The radical product of reaction (R6.24), CH_3, becomes involved in subsequent reactions that oxidise NO to NO_2 and regenerate OH radicals simultaneously. Methane oxidation proceeds thus:

$$CH_3 + O_2 \rightarrow CH_3O_2 \tag{R6.27}$$

$$CH_3O_2 + NO \rightarrow CH_3O + NO_2 \tag{R6.28}$$

$$CH_3O + O_2 \rightarrow HCHO + HO_2 \tag{R6.29}$$

$$HO_2 + NO \rightarrow NO_2 + OH \tag{R6.30}$$

Fig. 6.10. The concentration of gases in photochemical smog as a fuunction of time. This follows the classical representation of Leighton (1961).

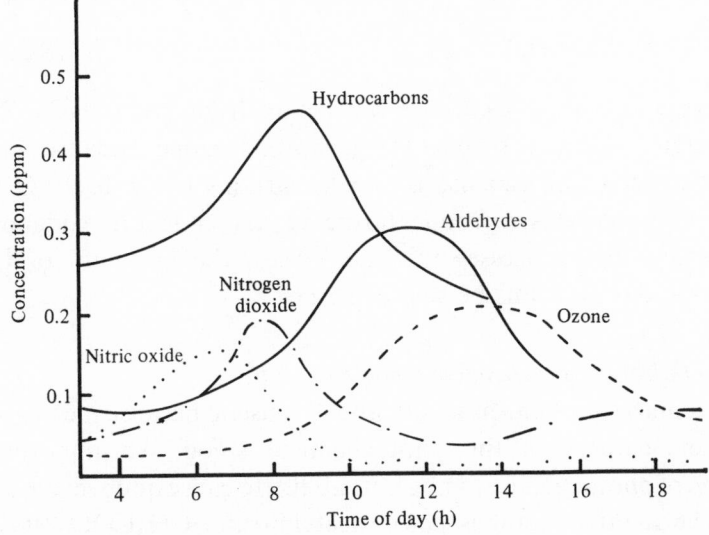

HO_2 can also react photochemically or with O_3, H or O to regenerate hydroxyl radicals. Formaldehyde (HCHO) can photodissociate into H and HCO or react with O_2 to give HO_2 and CO. The reactions (R6.24) and (R6.27)–(R6.30) can be summed to show the importance of the presence of hydrocarbons in generating nitrogen dioxide in photochemical smog:

$$CH_4 + 2O_2 + 2NO \rightarrow H_2O + HCHO + 2NO_2 \qquad (R6.31)$$

which indicates that the oxidation of nitric oxide has not used ozone. This was not true with the pseudo-equilibrium situation described by reactions (R6.21)–(R6.23). Thus the presence of an alkane such as methane in the polluted urban air provides a way in which nitric oxide can be oxidised without consuming ozone.

The attack of OH on acetaldehyde yields CH_3CO (R6.25), which oxidises along the following path:

$$CH_3CO + O_2 \rightarrow CH_3COO_2 \qquad (R6.32)$$

$$CH_3COO_2 + NO \rightarrow NO_2 + CH_3CO_2 \qquad (R6.33)$$

$$CH_3CO_2 \rightarrow CH_3 + CO_2 \qquad (R6.34)$$

The CH_3 now enters the reaction pathway at (R6.27).

The atomic hydrogen produced in reaction (R6.26) or from the photodissociation of formaldehyde (Reaction 5a in Table 3.2) can react with HO_2 to produce two OH radicals that can in turn initiate further attack on organic compounds, or it can form a hydroperoxyl radical:

$$H + O_2 + M \rightarrow HO_2 + M \qquad (R6.35)$$

Thus, aldehydes also provide effective ways of oxidising NO to NO_2. We can now imagine this as a scheme for generating ozone, because, in a polluted atmosphere, we can use the hydrocarbons to oxidise NO to NO_2. This NO_2 photolyses to produce further O_3 and NO for re-oxidation. While there are loss processes in these cycles, the build up of O_3 throughout the day can thus be well explained.

Other products from photochemical smog reactions
The photochemical mechanism set out above is useful because it can also explain other features of the photochemical smog. An important characteristic of photochemical smog is its ability to cause quite severe eye irritation. The major irritant is peroxyacetylnitrate ($CH_3COO_2NO_2$),

which is often abbreviated to PAN. Reaction (R6.32) showed how the CH_3COO_2 radical formed. This radical can react with NO_2:

$$CH_3COO_2 + NO_2 \rightarrow CH_3COO_2NO_2 \qquad \text{(R6.36)}$$

This compound is the principal member of a homologous series of similar nitrated compounds, which include higher peroxyalkyl and some aromatic forms. Modern motor fuels have often tended to take on an increasingly aromatic character that has appeared to have accompanied the removal of lead in fuel. Compounds such as benzene and toluene are released in large quantities by vehicles and the peroxybenzoyl nitrate that is formed is a particularly potent eye irritant. The PAN class of compounds can often be detected in the remote atmosphere if there are large pollution sources upwind.

Although some nitrogen oxide in the polluted atmosphere will end up as PAN type compounds through reactions of the type (R6.36), other important reactions are possible:

(i) oxidation of NO_2 to NO_3:

$$NO_2 + O_3 \rightarrow NO_3 + O_2 \qquad \text{(R6.36)}$$

which will subsequently react with NO_x

$$NO_3 + NO_2 \rightarrow N_2O_5 \qquad \text{(R6.37)}$$

$$NO_3 + NO \rightarrow 2NO_2 \qquad \text{(R6.38)}$$

(ii) reaction with OH:

$$NO + OH \rightarrow HNO_2 \qquad \text{(R6.39)}$$

$$NO_2 + OH \rightarrow HNO_3 \qquad \text{(R6.40)}$$

(iii) reaction between NO and NO_2:

$$NO + NO_2 + H_2O \rightarrow 2HNO_2 \qquad \text{(R6.41)}$$

An end product of these oxidation reactions is nitrous or nitric acid. Nitrous acid (HONO or HNO_2) is liable to undergo photochemical dissociation according to the reaction

$$HNO_2 + h\nu \text{ (less than 400 nm)} \rightarrow NO + OH \qquad \text{(R6.42)}$$

and so represents a source of OH. The formaldehyde present in the polluted atmosphere (from (R6.29)) constitutes an important source of

atomic hydrogen and therefore of hydroperoxyl and hydroxyl radicals
(when the two reactions below are followed by (R6.30) or (R6.35)):

$$HCHO + hv \text{ (less than 370 nm)} \rightarrow H + HCO \qquad (R6.43)$$

$$HCHO + hv \text{ (less than 370 nm)} \rightarrow H_2 + CO \qquad (R6.44)$$

Hydrogen peroxide may be produced through two reactions:

$$M + OH + OH \rightarrow H_2O_2 + M \qquad (R6.45)$$

$$HO_2 + HO_2 \rightarrow H_2O_2 + O_2 \qquad (R6.46)$$

Sulphur dioxide can also be oxidised under photochemical conditions,
but the sulphur–oxygen bond is very strong so that sulphur dioxide
cannot undergo the photodissociation with which we are familiar for
nitrogen dioxide (R6.21). The oxidation process involves the hydroxyl
radical and may be represented by

$$OH + SO_2 \rightarrow HSO_3 \qquad (R6.47)$$

$$HSO_3 + O_2 \rightarrow HSO_5 \qquad (R6.48)$$

$$HSO_5 \rightarrow HO_2 + SO_3 \qquad (R6.49)$$

$$SO_3 + H_2O \rightarrow H_2SO_4 \qquad (R6.50)$$

although (R6.48) and (R6.49) probably occur as a single reaction:

$$HSO_3 + O_2 \rightarrow HO_2 + SO_3 \qquad (R6.51)$$

Larger organic molecules
In the opening discussion of the photochemistry of the polluted atmosphere,
we restricted the organic chemistry to methane and acetaldehyde.
However, it should be emphasised that these are only model compounds.
Smog chemistry invariably involves much more complex molecules.
Although the alkanes, especially the larger ones, are readily attacked by
the OH radical in the atmosphere, other classes of organic compounds are
even more active in Los Angeles-type atmospheres. It is often useful to
describe organic compounds in terms of their *photochemical ozone
creation potential* (POCP). The values are chiefly related to the rate of
attack by OH and then the scale set such that ethylene, which is a strong
ozone producer, takes the value 100. We should note that accurate
estimates of the POCP require more than a simple measure of the rate of
OH attack. This is because other species in the atmosphere can be

Fig. 6.11. Monthly average daily maximum ozone concentrations at the University of Mexico monitoring station (from Bravo *et al.* (1991)).

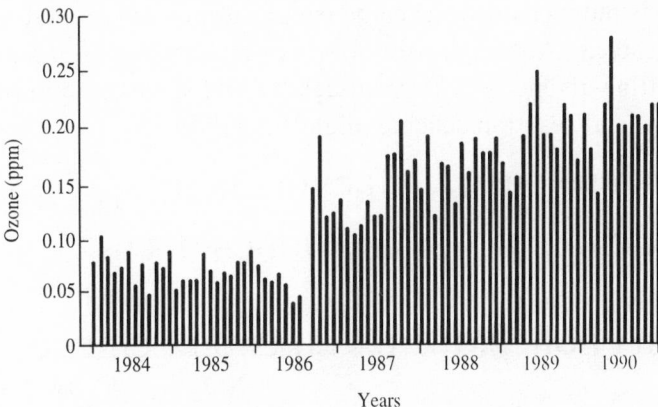

involved in smog reactions that will produce or consume ozone. The potential of aromatic compounds, such as toluene and benzene, to produce smog is clear from their high POCP (see Table 6.2).

The introduction of unleaded fuels can induce more serious photochemical smogs because of the enhanced degree of reactivity (often as increased aromaticity) of the emissions. This is particularly true if the change is not accompanied by the introduction of catalytic converters. Dramatic changes took place in Mexico City after the introduction of a new fuel in September 1986. There was a rapid increase in ozone concentrations immediately after the fuel change (see Fig. 6.11). This new fuel was unleaded because of a desire to reduce the lead concentrations in the atmosphere of the city. Petroleos Méxicanos has been reluctant to release information on the formulation of its gasoline, but it has been widely assumed that the octane rating of the new unleaded fuel was maintained through the addition of more reactive hydrocarbons. The experience in

Table 6.2. *Photochemical ozone creation potential (POCP) for various VOCs*

Gas	POCP
Methane	0.7
Ethane	4.0
Ethene	100
n-Butane	41
Benzene	19
Toluene	56

Mexico City has emphasised the sensitivity of the urban photochemistry to fuel formulation.

Alkenes, such as butene, are attacked by ozone, atomic oxygen, $O(^3P)$ or the hydroxyl radical. Although ozonolysis is very familiar to organic chemists for splitting alkene molecules, attack by OH is common in the atmosphere. A typical reaction scheme might be written

$$OH + CH_3CH{=}CHCH_3 \rightarrow CH_3CHOHCHCH_3 \qquad (R6.52)$$

$$O_2 + CH_3CHOHCHCH_3 \rightarrow CH_3CHOHCH(O_2)CH_3$$
$$(R6.53)$$

$$NO + CH_3CHOHCH(O_2)CH_3 \rightarrow CH_3CHOH + CH_3CHO$$
$$+ NO_2 \qquad (R6.54)$$

The reaction scheme shows the degradation of larger organic molecules into smaller, more oxidised ones.

6.6   Heterogeneous reactions – particulate materials and rainfall

Photochemical smog yields aerosols that give rise to the visual obscurity that is associated with photochemical pollution. However, the high opacity of smog gives an exaggerated impression of the amount of particulate material produced. It is estimated that as little as 5% of the pollutants present in photochemical smog could be converted to suspended particulate materials. The material that forms the condensed phase of smog consists of both inorganic and organic materials. The inorganic part includes metal oxides and the salts of acids produced within the urban air. Acids themselves (sulphuric and nitric) are present in association with solid particles, as salts or acidic droplets. Acidic droplets can react rapidly with atmospheric ammonia to form ammonium sulphate and nitrate aerosols that can be important causes of the reduction in visibility.

In the more traditional polluted winter atmospheres of coal burning cities liquid droplets are an important site for reactions such as the oxidation of sulphur dioxide to sulphuric acid. This oxidation reaction was discussed in Section 5.5, but in polluted atmospheres the droplet reactions can be faster because the concentrations of metal ion catalysts, or hydrogen peroxide and ozone (in photochemical smogs), may be much higher. In particular, metal ions leached from particles released into the air through anthropogenic activity can end up in urban aerosols or rain (e.g. Table 6.3). Particles of coal and refuse derived fuel ash are likely to be ready sources of catalytic metals for droplets. It is possible that oxidation rates for sulphur dioxide further increase through the dissolution of materials such as calcium oxide (which is present in high concentrations in coal fly ash), which will make the droplet alkaline:

$$CaO + H_2O \rightarrow Ca^{2+} + 2OH^- \tag{R6.55}$$

This allows larger amounts of SO_2 to dissolve and increase the rate of catalytic oxidation. Alternatively the dissolution of ammonia from a polluted atmosphere could also increase pH and enhance both dissolution and oxidation of sulphur dioxide.

Another heterogeneous route for oxidation of sulphur dioxide is absorption of the gas onto solid surfaces followed by subsequent oxidation. The difficulty with such a mechanism is that it requires some method for the surface to be 'cleaned'. This is required because the area of the particulate material, even in a polluted atmosphere, is small and would soon be covered. However, if the particles were wet then the mechanism could be effective. For instance, the oxidation of SO_2 in solution in the presence of dissolved NO_2 and suspended soot proceeds more rapidly with the soot particles acting as a catalyst.

Table 6.3. *Composition of rain* (μmoll^{-1}) *at some urban sites*

	Athens[a]	Los Angeles[b]	Montréal[c]
Calcium	49	4.2	
Magnesium	13	2.7	
Potassium	7	1.6	
Sodium	87	21.0	
Ammonium	28	16.8	
Sulphate	50	10.7	
Nitrate	28	16.6	
Chloride	100	21.4	
Hydrogen peroxide		4.4	
Formate		12.4	
Acetate		4.1	
Formaldehyde		3.3	
Acetaldehyde		0.2	
Iron			1.17
Aluminium			0.45
Zinc			0.29
Manganese			0.10
Lead			0.01

[a]Smirnioudi and Siskos (1992).
[b]Sakugawa et al. (1993).
[c]Poissant et al. (1994).

Even where metals are not catalysts for reactions, they are of interest because of their toxicity. Metals are present, for the most part, as solids rather than in solution. Their distribution among the particles of the urban atmosphere depends on the way in which they were emitted. Volatile metals (i.e. those with boiling points $<1500\,°C$) are vaporised during combustion processes and subsequently condense on the surfaces of particles. Such metals are thus concentrated on the surfaces of particles and readily available for biological interactions. This is particularly relevant when we consider that volatile metals such as cadmium and lead are very toxic.

Lead, because of its use in automotive fuels, is a special metal in the urban environment. Although leaded fuels are increasingly being phased out, they will be used in some countries for many years. The combustion process oxidises over 90% of the tetra-alkylated lead in fuels, so only a small fraction finds its way into the atmosphere in the organic form. The rest is converted to lead oxides that react with the HCl and HBr in exhaust gases to produce a range of halide salts, the most important of which are $PbBr_2$ and $PbBrCl$. There is usually excess HCl in exhaust gases, so this can react with atmospheric ammonia and form a mixed salt, $2PbBrCl·NH_4Cl$. Under conditions where there is sufficient ambient sulphuric acid or ammonium bisulphate, bromine and chlorine can de-gas from the aerosol into the atmosphere as HCl or HBr. This will ultimately give lead ammonium sulphates such as $PbSO_4·(NH_4)_2SO_4$.

However, the process is more complicated than simple displacement of a volatile hydrogen halide from a salt of the kind familiar in the marine aerosol shown in reaction (R5.20). Hydrogen bromide is very much more soluble than hydrogen chloride, so one might expect ageing lead salts in the atmosphere to become enriched in bromide. However the opposite is observed and bromide is lost until a stable salt configuration is reached, when the ratio Br/Cl approaches unity. This may well explain the relatively high abundance of PbBrCl in urban atmospheres. Lead dibromide particles darken in sunlight with the loss of halogen atoms, but the photochemical changes are not a major influence in controlling the composition of atmospheric lead particles.

Organic solids have long been thought to be produced in polluted urban air, and this notion has doubtless been aided by the ease with which particulate material arises through the oxidation of naturally produced terpenes. The oxidative nature of atmospheric photochemistry would suggest that oxidised compounds should be found in smogs. Measurements show that long-chain aldehydes and carboxylic and dicarboxylic acids are frequent particulate components of polluted air.

There are many primary pollutants that appear as organic particles. Very often although they exist in the form of particles because they have high boiling points, a fraction can exist in the gas phase too (see Section 4.4). A most important class of these organic pollutants is the polynuclear aromatic hydrocarbons (PAHs) of which benzo(a)pyrene is best known. It was produced in large quantities from domestic coal fires, but today, although coal is used less, it is still an important source (Fig. 6.12). The carcinogenic properties of many of these PAHs are the major reason for the concern over their presence in the atmosphere. More recently there have been studies of nitrated forms of PAHs, because of the likelihood that these are especially potent carcinogens. The increasing diesel fleet is a possible source of these. Chlorinated PAHs, such as the chloro- and dichloropyrenes, have now also been detected in urban air.

The declining use of coal in urban areas initially offers some improvements in the soiling qualities of air because of decreases in the particulate load. However, in many European cities these gains have tended to be lost with the increasing production of particles from diesel vehicles. Diesel particulate material is very effective at soiling the urban fabric. Particles, from this source, are small, sticky and extremely black entities and have become the principal contemporary cause of the blackening of buildings.

Urban rainfall is dominated by the dissolved constituents of the principal air pollutants. In particular sulphates and nitrates dominate over the marine salts that would be typical of remote coastal locations. Table 6.3 shows that, besides the expected elevation in sulphate and

Fig. 6.12. The sourcs of PAH in the UK (1985), which amount in total to 176 tonnes a^{-1}.

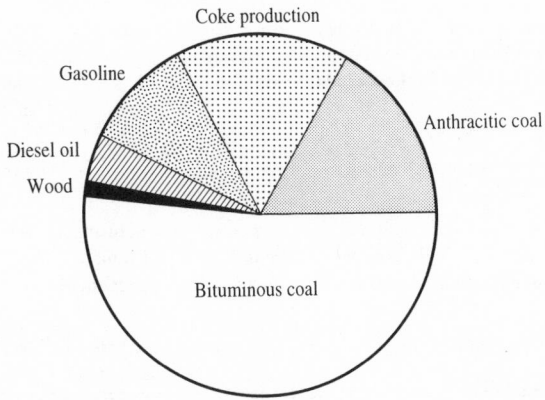

nitrate concentrations, rainfall in Athens has high concentrations of calcium from soil dust. This can be compared with the high calcium concentrations that were noted for the Pyrenees (Table 5.4)

6.7   Air pollution meteorology

Meteorology is not a source of air pollution in the same way as the sources discussed in the previous sections, yet pollution of the atmosphere would be a less serious problem were it not for the vicissitudes of the weather. Often the mobility of the atmosphere is high enough to disperse air pollutants rapidly and prevent the most serious consequences. If we examine the characteristics of the worst air pollution disasters, listed in Table 6.4, we find certain similarities. In particular, we see the importance of calm conditions, which are often found to be associated with anticyclonic conditions, fogs and inversions. Excepting the Pozo Rica accident, the importance of valleys in air pollution events is also evident. The Pozo Rica incident occurred in a coastal setting. Both this and the recent disaster at Bhopal occurred at night, a time when winds are generally

Table 6.4. *Air pollution disasters*

	Meuse Valley, Belgium 1930	Donora, Pennsylvania 1948	Pozo Rico, Mexico 1950	London, UK 1952	Bhopal, India 1984	London, UK 1991
Mortality and morbidity						
Deaths 60		15	22	4000	2500?	160
Ill 6000		5900		>20000	10000?	
Age groups affected						
	elderly	elderly	all ages	elderly at first		patients with respiratory illness
Weather						
	anticyclonic inversion and fog	anticyclonic inversion and fog	nocturnal inversion, low winds	anticyclonic inversion and fog	nocturnal low winds(?)	anticyclonic inversion and fog as in 1952
Geographical setting						
	river valley	river valley	coastal	river plain		river valley
Sources						
	steel and zinc manufacture	steel and zinc manufacture	sulphur recovery (accident)	domestic coal burning	fracturing of tank (accident)	vehicles
Pollutants						
	SO_2, smoke	SO_2, smoke	H_2O	SO_2, smoke	methyl-isocyanide	NO_2, particles

lighter. Coastal environments also have their own characteristic meteorology that can enhance the accumulation of air pollutants. The London smog of 1991 shows that the deaths attributable to these extreme conditions have declined considerably with lowered emissions of air pollutants in cities. In the smog of 1952, deaths were blamed on the high concentrations of smoke and SO_2, while in the modern winter smog these pollutants were much reduced and NO_2 and fine diesel particles are more likely culprits. Future reductions in death rate will need improvements in maximum concentrations of particles or nitrogen oxides.

The temperature structure of the atmosphere controls the vertical movement of pollutants, as noted in Section 1.4. Decreases in temperature with altitude are really only an 'averaged' description. If the actual change in temperature in the ambient air is greater than the lapse rate, then a rising pocket of air will find itself at a higher temperature than the surrounding air. This means that it will have a greater tendency to rise. This condition is termed 'unstable'. If the environmental lapse rate is the same as that expected due to adiabatic expansion then the condition will be neutral. Under conditions in which the environment cools less rapidly with height than the adiabatic lapse rate an inversion is said to have formed.

Inversions are important controls on dispersion because they inhibit vertical transport. Radiation inversions form at night when the ground cools more rapidly than the air. During the day such inversions break up as the sun warms the ground. This can lead to more complex thermal structures in the lower few hundred metres of the atmosphere. These can be responsible for the great variety in the shapes of factory chimney plumes that are observed (see Fig. 6.13).

Dispersion of plumes from point sources
The dispersion of pollutants from a tall chimney is an important process and it is often necessary to predict how effectively it occurs. This dispersion is complex, but is often modelled, crudely, by assuming that the pollutants spread, in a Gaussian manner, as they move outwards from the plume axis (Fig. 6.14). Usually the gases within the stack are hot so that their point of origin is imagined to be somewhat above that of the chimney top. The other problem that has to be overcome is the fact that chimneys are often not very high so that the plume soon hits the ground. This is done by incorporating a reflection at ground level into the equations. There are many difficulties in producing a usable equation to describe the dispersion accurately under all conditions. The equation below is one of a variety available to calculate plume dispersal:

Fig. 6.13. Commonly observed behaviour of factory smoke plumes. Views shown are at right angles to the wind direction and the plunes are also given in cross section. (*a*) Coning. This occurs under near-neutral conditions, which lead to approximately equal dispersion in both the horizontal (cross-wind) and vertical directions. (*b*) Lofting. This occurs when emission is just above an inverson layer. (*c*) Looping. This occurs under unstable conditions. (*d*) Fanning. This occurs under very stable conditions (i.e. an inversion), which leads to much horizontal dispersion, but little vertical spread. (*e*) Fumigation. This is caused by emission just below an inversion layer.

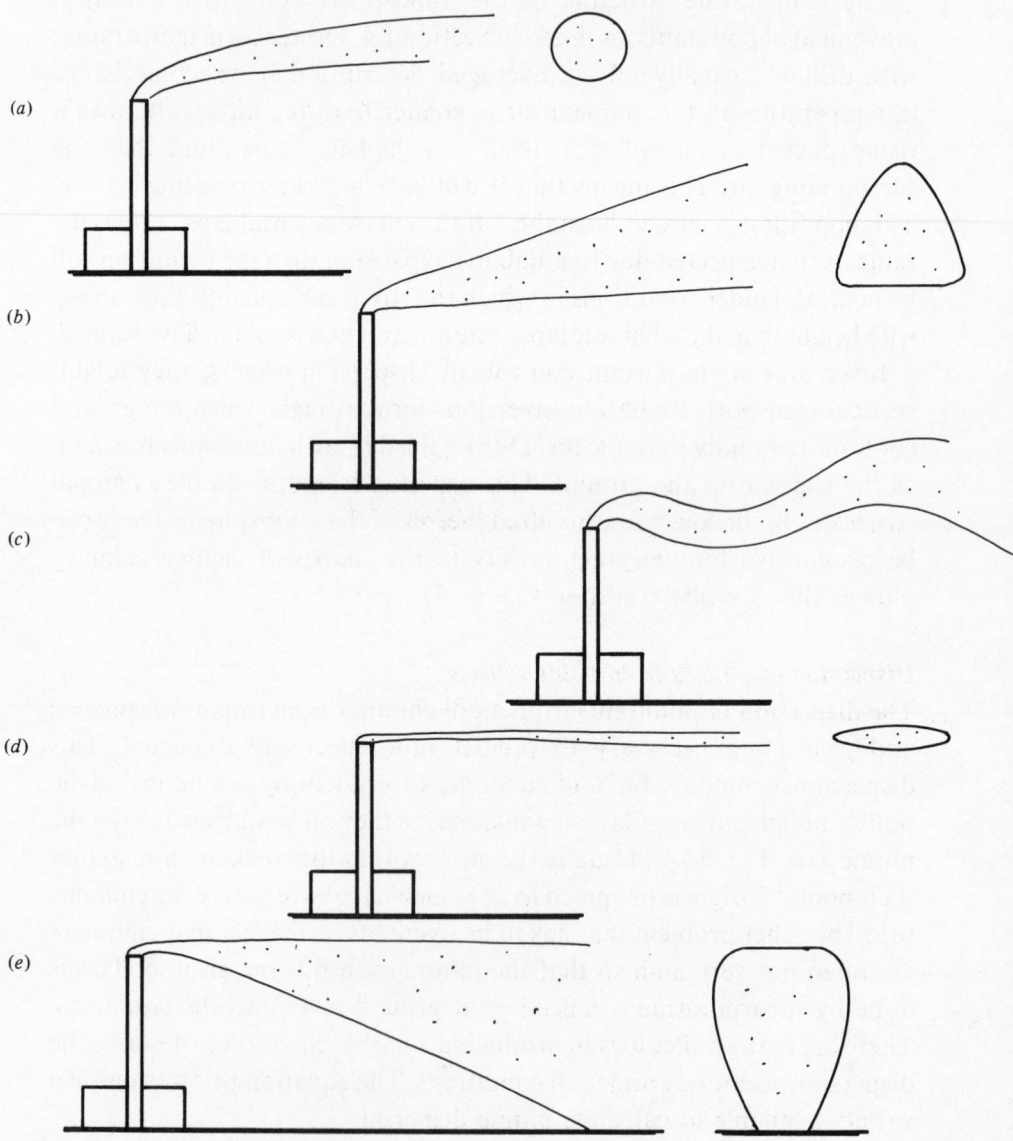

$$c_{x,y,z} = [Q/(2\pi U s_y s_z)] \exp [- (y/s_y)^2/2]$$
$$\times \{\exp\{ - [(z - H/s_z]^2/2\} + \exp\{ - [(z + H)/s_z]^2/2\}\} \quad (6.4)$$

where s_y and s_z are the diffusions in y and z directions and may be thought of as the standard deviation of concentration along the relevant axes:

> Q is the source strength (g s^{-1})
>
> x,y and z are distances along the axes defined in Fig. 6.14 (m)
>
> c is the concentration at position x,y,z (g m^{-3})
>
> U is the average wind speed (m s^{-1})
>
> H is the effective height of emission. (i.e. the sum of the chimney height and the plume rise).

It is difficult to give average values for the various coefficients required by this formulation. Solutions for restricted sets of conditions may be found in some books (e.g. Henderson-Sellers, (1984) and Seinfeld (1986)). We can see that the first term of the equation in square brackets represents the decrease of concentration with distance from the source. However, the exponential terms can increase with distance. This gives rise to the maximum in ground level concentration that occurs some distance downwind of the chimney. In Fig. 6.15 are shown some results from the application of the dispersion equation.

Fig. 6.14. Axes adopted for describing a plume from an elevated point source with plume rise Δh. A warm plume is sufficiently bouyant to rise some distance above the top of the chimney. This means that the plume behaves as if its source were above the chimney.

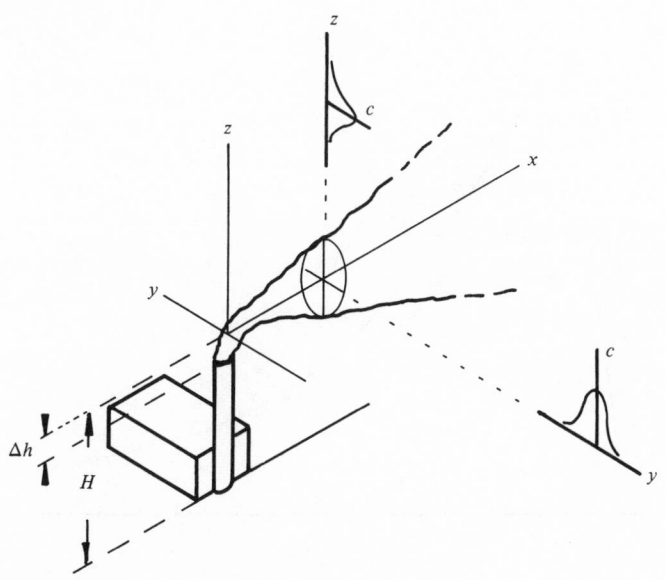

Calculations of this type are used to provide information during planning exercises. However, these formulations present many difficulties. The use of diffusion or dispersion coefficients from short-term experiments is not particularly suitable for longer time scales. Gaussian models have to be modified to account for chemical reactions and absorption at ground level. They do not allow for complex topography and vertical profiles of wind velocity and air temperature. The effect of topography can be particularly dramatic. Plumes may fumigate areas in the downwash of air as it crosses a ridge or an obstacle such as a nearby building (see Fig. 6.16). There have been many studies of the dispersion of air pollutants in complex terrain because of its great importance in assessing the risk of serious pollution from the construction of new sources.

Dense gas dispersion
In the cases discussed above, the pollutants are considered to be neutral or slightly buoyant in the atmosphere. In recent years industrial accidents

Fig. 6.15. Ground level concentrations of sulphur dioxide along the x-axis of a plume. Calculations are presented for two different emission heights (250 and 300 m). In this calculation, a wind speed of $7.5\,\mathrm{m\,s}^{-1}$ at the plume height has been used under stability class C. This stability condition is often found at moderate wind speeds in sunny conditions. The total rate of emission was set at 120 tonnes of sulphur per day. Note how the higher chimney height leads to a lower peak ground level concentration and that this peak lies further downwind of the source.

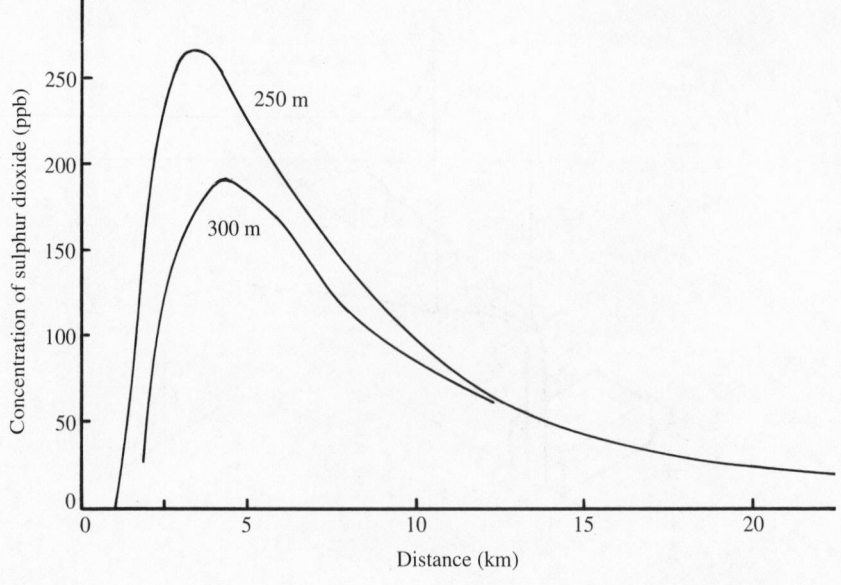

have shown that dense gases may disperse in a way that is not described by the classical approaches. There are two well-known cases of dense gas dispersion: (i) venting of material at high pressure and (ii) gravity-dominated dispersion of dense vapour clouds.

The first of these is typical of industrial accidents caused by overpressure in a reaction vessel. This leads to the loss of material through a relief valve or, in the more serious cases, total failure of the pressure vessel. The best known recent accident of this type was that at Seveso. In this accident the exit velocity of gas through a ruptured bursting disc was estimated to have been $270 \, \text{m s}^{-1}$. This high-velocity vertical jet was quite different from the gentle buoyant rise of a chimney plume. Such jets take on a conical shape with a roughly $10°$ half angle. Descriptions of the Seveso release say that it looked like a vertical ice cream cone. The plume then bent over and was advected by the wind, gradually depositing material in the lee.

Dense vapour clouds are commonly associated with liquid gas spills. While theoretically it would be thought that a large spill of liquid gas would lead only to partial vaporisation, the entrainment of air causes vaporisation to be almost complete (see Fig. 6.17). The cloud of vapour rises into a right cylinder with the dense gas at a concentration of about 10%. This cylinder then slumps because of its high density. The leading edge of the expanding cloud entrains more air that causes it to warm up further. Finally, when the gas becomes warm, it can rise from the ground and undergo the familiar diffusion and advection. In a 20 tonne liquid natural gas release an initial cylinder of 100 m diameter would be expected. Within 30 s it would slump to some half a kilometre across. The gas velocities involved in such an accident are high enough to make escape difficult.

Fig. 6.16. Downwash of a plume in the lee of a ridge.

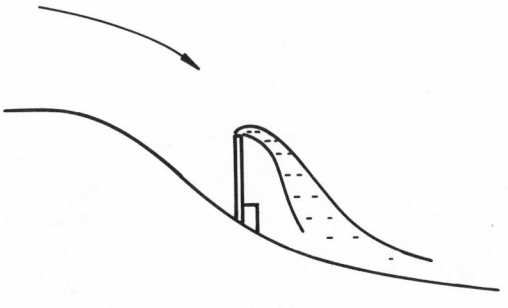

Larger scale dispersion

In cities, inversions often serve to trap the pollutants at a low level. Radiation inversions of a few tens of metres depth were typical of the conditions that aided the formation of classical smogs in London. There are also other types of inversion. Cold air associated with an advancing weather front can slide underneath warm air to produce a frontal inversion. Cool mountain air can drain down into warmer valleys at night producing an advective inversion. Larger scale movements of air may also give rise to inversions. Subsidence inversions arise through large scale downward movement of air associated with the convective cells responsible for the global circulation of the Earth's atmosphere. This type of inversion is often responsible for the high stability of the air in the Los Angeles basin. Here the subsidence inversion, which can be about 1000 m thick, serves to trap air within the basin. The situation is exacerbated by the presence of a wall of mountains to the east, whose summits are higher than the inversion, and by the presence of sea breezes (see Fig. 6.18).

Sea breezes are regular diurnal phenomena. During the day, air warmed by a rise in temperature of land surfaces rises vertically. This draws a sea breeze off the oceans. At night the reverse occurs and a land breeze runs out to sea. It is not unknown for these breezes to remove pollution from a coastal city during the night only to deposit it back there

Fig. 6.17. The stages in dispersion of a spill of liquified gas: (*a*) The spill of liquid, (*b*) vaporisation and (*c*) slumping.

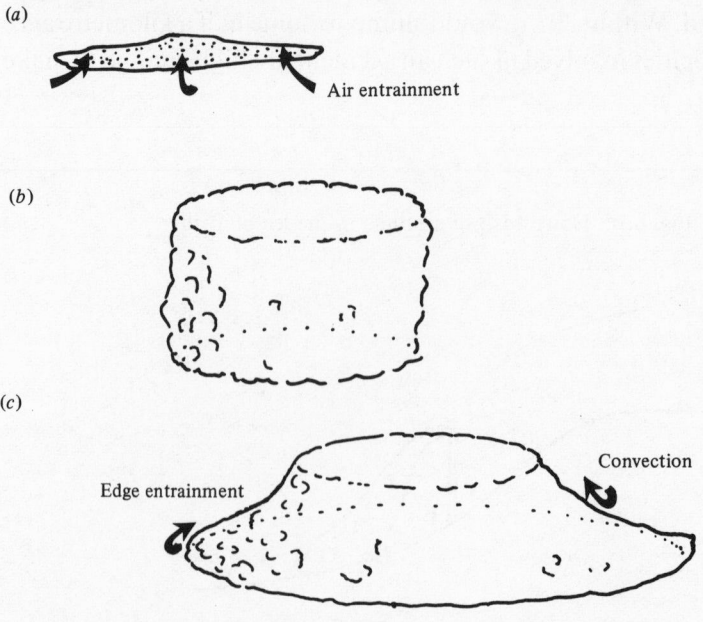

next day. In very hot locations, where the land does not have the chance to cool at night, the sea breeze can be almost continuous. It could take the pollution away only to bring it back through a complete circuit of the sea breeze cell.

Valleys also have characteristic diurnal winds. The cycle of anabatic and katabatic winds moves air up and down the valley. The sides of the valley constrain sideways movement, while the existence of an advective inversion may prevent escape from the top. Thus polluted air may run up and down the valley in a diurnal cycle. A well-known occurrence of pollution in such a situation is that of the smelters of Trail on the Columbia River. The pollution from the smelters drifted down the valley and across the USA–Canada border into Montana, where it caused a loss of crops.

The complexities of the dispersion of air pollutants on a regional scale make modelling difficult. It is not possible to reduce such models to simple equations. Generally such situations are examined by following the movement of many individual air parcels. The enormous number of calculations requires the use of a computer. For urban regions several simple models have been suggested that treat the city as a single parcel of air and enable very crude estimates of the mean pollutant concentration to be made, e.g.

$$c = Q/(0.45Ud^{1.68}) \tag{6.5}$$

where Q is the amount of pollutant emitted by the city in $g\,s^{-1}$ (which may be estimated from economic or demographic data), U is the mean wind speed $(m\,s^{-1})$ for the period under consideration and d is the diameter of the city in metres. The estimated concentration of the pollutant takes the units $g\,m^{-3}$. For individual cities it is possible to correlate average pollution levels with various meteorological parameters such as temperature, wind speed and wind direction. Unfortunately the correlations are often not very good, but sometimes, at least, they offer an opportunity to utilise

Fig. 6.18. Pollutants trapped by a mountain barrier, sea breeze and an inversion layer. This idealises the situation found along the west coast of North America.

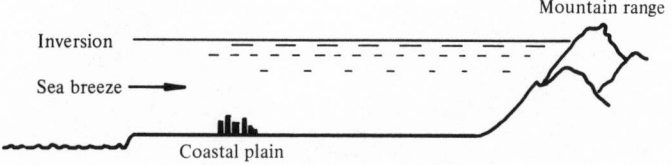

weather forecasts to make air pollution forecasts. Forecasts are more effective if they try to model in a fairly accurate way both future emissions and meteorological dispersion patterns, together with the chemistry. Such forecasting techniques are particularly important where it is necessary to issue public alerts about episodes of poor air quality. These forecasts would also allow pre-planned reductions in emissions from industry and transport to prevent a serious air pollution episode.

6.8   Further reading

Henderson-Sellers, B. (1984) *Pollution of Our Atmosphere*, Adam Hilger, Bristol.

Seinfeld, J. H. (1986) *Air Pollution*, John Wiley and Sons, New York.

7

○ ○

Air pollution and its effects

Air pollutants are trace components of the atmosphere that are present at unexpectedly high concentrations. However, it is usual to call them pollutants only when they exhibit some detrimental effect. Small helium leaks might well raise the helium concentration in a local area when large quantities are in use, but the helium would be unlikely to be regarded as an air pollutant until some effect was apparent. Another difficulty in discussions of pollutants is the position of 'natural pollutants'. It is often hard to know whether to class emissions from volcanoes or swamps as pollution yet these can have environmental consequences just as serious as the more conventional anthropogenic emissions. It seems that we are likely to refer to natural emissions as pollutants when they affect our lives.

Air pollution occurs almost everywhere and has a long history. In this chapter we shall look at some situations in which air pollution occurs, but should remember that these are merely illustrative. With the increasing complexity of our society, an enormous range of pollutant compounds is found in the atmosphere, so here we will cover the frequent types of air pollution.

7.1 Urban air

The pollution of urban air is familiar to any city dweller. It can often be seen and smelled. This is particularly noticeable because of the high density of emission within cities. Classical writers tell us that cities of Ancient Rome and Greece were polluted. However, it was the transition to coal that began in thirteenth century medieval England and had spread worldwide by the nineteenth century that initiated a form of air pollution that has characterised the atmospheres of cities until recent times. The use of coal, both in domestic and in industrial grates, caused cities to be very smoky places. High concentrations of sulphur dioxide accompanied the smoke and for a long time these two pollutants have been synonymous with urban air pollution or smog. The term smog was derived from the words '*smoke*' and '*fog*'. Such smogs were prevalent in coal-burning cities under calm winter conditions when the smoke particles acted as condensation nuclei for fog.

Primary pollutants were found at the greatest concentrations near the centres of cities. Today there are polluting activities with a unique character, such as the specialised solvents used in the electronic industry. An industry like this can generate high concentrations of the solvent vapour in the air near the site of release. The inhomogeneous distribution of pollutants within a city causes problems when one wants to specify the pollutant concentration of a given city. It will depend on where it is measured. It is necessary to describe carefully the location of the measurements within the city if the pollutant concentrations are to be meaningful. There is often much disagreement and political debate over the location of monitoring sites. Groups campaigning for cleaner air will often press for them to be located in highly polluted places, perhaps at the roadside or along a factory perimeter. Scientists interested in long term changes in concentration or exposure of the urban population in general will wish for sites (such as pedestrian precincts or parks within the central city) that are less susceptible to highly localised influences.

The problems of describing pollutant concentration do not end once the monitoring site has been carefully described. Urban pollutants do not remain at a constant concentration in the atmosphere. The concentrations are often highly variable so that a particular measurement can easily represent a freak high or low value. Because of this (and also because of problems in making measurements very rapidly), concentrations are generally averaged over some period. One minute, five minutes, an hour, a day and the year are popular averaging times. As one would imagine, the longer the averaging time the more likely the value will represent pollution levels at the site. However, long averaging times cause the loss of valuable information about cycles and temporal changes. Yet long averages remain useful when we are concerned about the cumulative exposure to carcinogens such as benzene.

In Fig. 7.1 are shown some ways of representing pollutant concentration data. In Fig. 7.1(*a*) we see a plot of one-minute measurements as a function of time. The mean values for the ten-minute intervals, plotted as lines, show a lower degree of variation. When plotted as a frequency-histogram the one-minute measurements illustrate a skew (Fig. 7.1(*b*)). This is typical of air pollution data, which often approximate a log-normal distribution, so, if a frequency-histogram of the logarithm of the concentration were plotted, the distribution would be closer to symmetrical. The log-normal distribution becomes linear when plotted on probability axes as shown in Fig.7.1(c). Air pollution data are frequently plotted on such a diagram. Finally, Fig. 7.1(d) shows an arrow-head diagram. Here the data from diagrams such as Fig. 7.1(c) are taken and plotted as a function of the logarithm of the averaging time. The decrease in variation with averaging time, which has already been mentioned, is well illustrated by such a diagram. An important property of this figure is that it illustrates that, at almost every averaging time, the 30% frequency corresponds to the arithmetic mean concentration.

Classical smog

Traditionally, large cities, where the coal usage was high, tended to show annual mean SO_2 concentrations in excess of 0.1 ppm. The levels found in cities have dropped over the last thirty years as increasing attention has been paid to reduction of sulphur emissions. Today large cities probably expect an annual mean of half this value. In line with these improvements, the maximum daily values of SO_2 are also considerably lower. These dramatic improvements will probably not continue because attention has now shifted to other urban pollutants and further improvement would be increasingly hard to achieve. However, air pollution control measures have attempted to keep the SO_2 level from rising in the face of various fuel changes and increasing use of diesel fuel. Current methods of keeping SO_2 low try to limit the sulphur concentrations in fuel. Improvements in the SO_2 and smoke levels have not always been so marked in cities of the developing world, where a reliance on coal and rapid industrialisation has not helped the quality of urban air. There is a growing awareness, though, that the problems are acute in these locations and need attention.

Over a long period we can see that there have been substantial improvements in levels of the traditional smoke and sulphur dioxide pollutants. Taking Glasgow as an example (Fig. 7.2), there is a marked drop in SO_2 concentrations over time. In many larger cities of Britain or the eastern seaboard of North America this decline began perhaps at the beginning of the twentieth century. Such decreases pre-date the well known UK Clean Air Act of 1956, which is often seen as a turning point in modern air pollution control, and its counterpart in the USA of 1970. In Fig. 7.2(a) are shown SO_2 concentrations and in Fig 7.2(b), the average smoke concentrations. The smoke concentrations have traditionally been measured by aspiration of large volumes of air through filter papers. The

Fig. 7.1. Measurement of pollutant concentration. (*a*) Variation of pollutant concentration with time. Individual measurements vary more than the mean of ten measurements (horizontal lines). (*b*) Distribution of 100 measurements showing a log-normal form (five values were greater than 35 ppb). (*c*) A cumulative plot showing the percentage of measurements in which pollution concentration exceeded a given value. The nearly straight line confirms that the data shown in (*b*) were log-normally distributed. (*d*) The number of times, as a percentage, that pollutant concentration exceeded a give value as a function of sampling time. Note how the range of the measured concentrations decreases with increased averaging time.

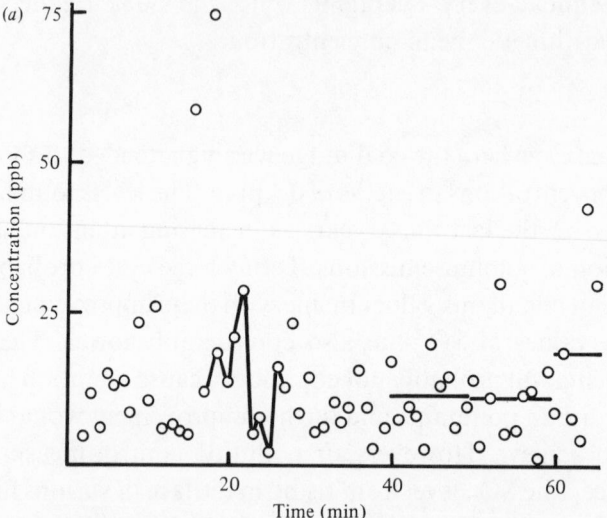

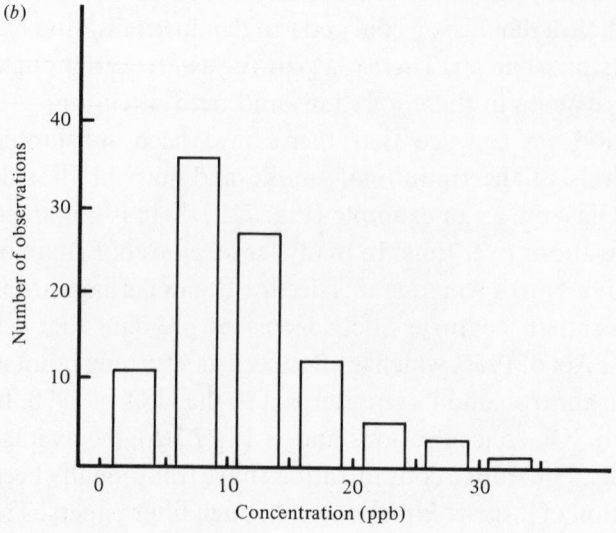

(c)

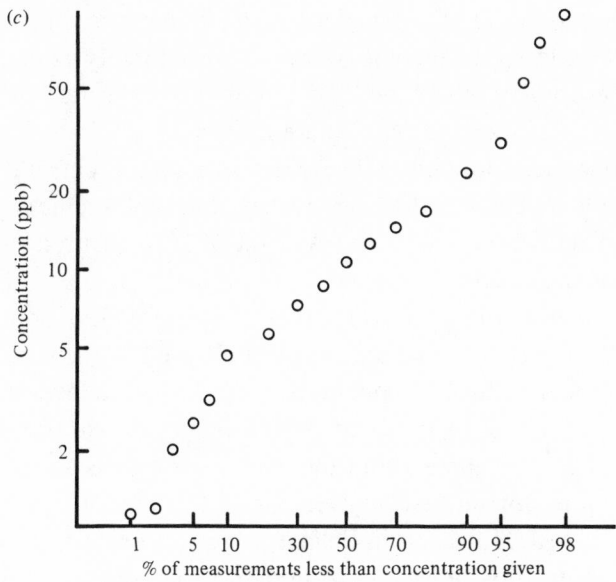

(d)

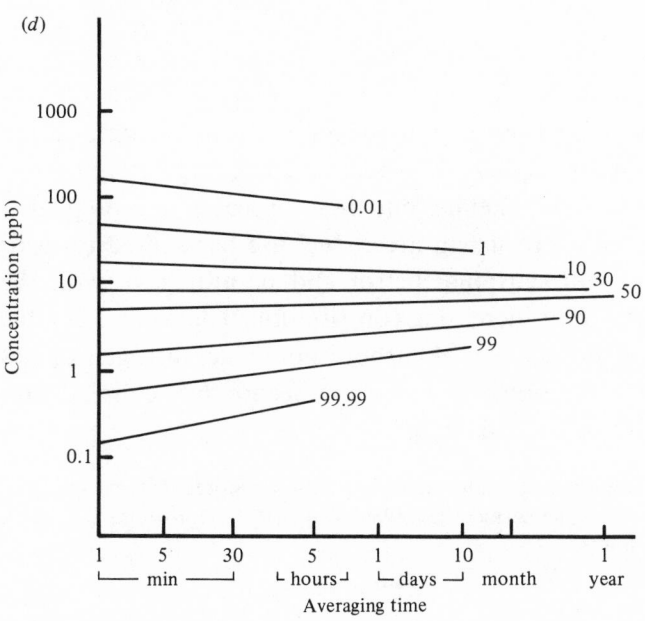

particulate material gathered on the filter paper can be measured in two ways: directly by determining the increase in weight or indirectly from the degree of soiling measured by reflectance. The amount of deposited solid is related to the level of suspended particulate material.

Many earlier improvements in sulphur dioxide and smoke concentrations were due to changing patterns of fuel use rather than enforcement of particular legislation. However, the Clean Air Act of 1956 ensured that changes due to burning cleaner fuels have continued. The Act also allowed local government to designate certain areas as Smokeless Zones. Within such zones the emission of smoke was not allowed. This usually required the use of smokeless fuels. Surprisingly it also helped to lower the sulphur dioxide concentrations in urban air, even though that was not the main purpose of the Act. This came about because the new fuels adopted under the Act had, quite fortuitously, lower concentrations of sulphur than those previously used. However, in Belfast, where natural gas is not readily available, sulphur dioxide concentrations remain comparatively high. More recent legislation in many countries allows the authorities to limit or reduce the sulphur or lead content of fuels.

Emissions from automobiles

With some exceptions, coal-burning no longer controls the composition of the modern urban atmosphere. It is still burnt in large quantities to produce electricity or refine metals, but these processes are frequently undertaken outside cities. In urban areas, fuel use has shifted towards liquid and gaseous hydrocarbons (petrol and natural gas), with the automobile making an ever increasing contribution. These more volatile fuels mean increased production of carbon monoxide, nitrogen oxides and volatile organic compounds in cities (see Section 6.5) coupled with the formation of photochemical smog.

Fig. 7.2. Concentration of sulphur dioxide (1 ppb = 2.86 µg m^{-3}) and smoke at a Glasgow site over the period 1970–1980 (DoE, 1992). Horizontal lines mark EU Directive limits.

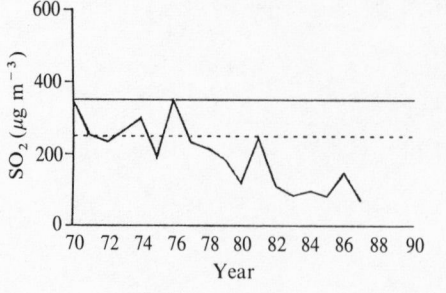
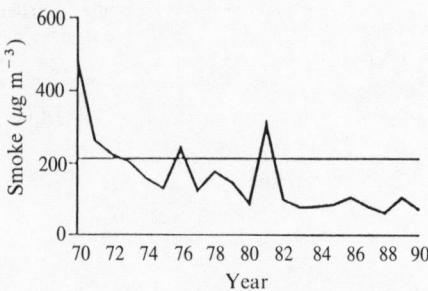

If we neglect carbon dioxide and water, the highest concentrations of pollutants in urban air are those of carbon monoxide. Because carbon monoxide originates in such large amounts from the internal combustion engine, its spatial and temporal distribution usually follows that of the traffic. Peak concentrations are often found early on weekday mornings and in the afternoons and evenings (Fig. 7.3). These correspond to periods of high traffic density. Carbon monoxide can be trapped within the urban canyons formed by the buildings that line city streets. In London, roads with heavy traffic often have carbon monoxide concentrations of 10–15 ppm and concentrations as high as 70 ppm have been measured in the Rotherhithe tunnel. Despite its high concentration, carbon monoxide currently receives much less attention than many other urban pollutants. This is probably because, in spite of its toxicity at high concentrations, its mode of action is fairly well understood. It seems to have few obvious effects on living things, which can metabolise it efficiently, at low concentrations, and structural materials are not affected. Enhanced concentrations, at high elevations, can arise from cars that have been tuned to operate at sea level. These can emit as much as four times as much carbon monoxide at 2500 m. This could pose a health risk in ski resorts as people are likely to be more sensitive to the pollutant at high altitudes or when exercising.

The concentration of lead in urban air has been high in the past because of the use of lead as an anti-knock agent in motor fuels. In London some

Fig. 7.3. Diurnal variation of the traffic-derived pollutants NO_x and CO (HMSO).

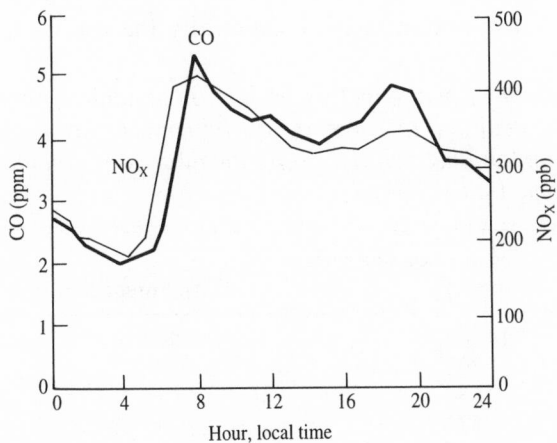

very busy streets sometimes reached concentrations of $3\,\mu g\,m^{-3}$. This exceeds the European Union guideline of $2\,\mu g\,m^{-3}$ and the US Environmental Protection Agency ambient air quality standard of $1.5\,\mu g\,m^{-3}$. The dust in London also had high concentrations of many heavy metals (Table 7.1). The abundance of lead was the result of vehicular emissions, but cadmium may also originate from cars, being found both in rubber and in sump oils. High enrichment of zinc and nickel compared with their crustal abundance (see Section 4.3 for a definition of enrichment) is in line with the fact that these two metals are enriched in oils (see Fig. 6.7). Zinc is added to oil as zinc dithiophosphate and zinc oxide is present in tyres.

Lead concentrations have declined in many countries over the past decades. This parallels the decline in emissions largely through the adoption of low-lead and unleaded fuels (see Fig. 6.8). Unleaded fuels are required by modern cars equipped with catalysts to lower exhaust emissions. However, perhaps the sharpest reductions in lead levels have come about when governments have decided to change fuel composition and lower the amount of alkyl lead permitted. These reductions can take place rapidly because they involve changes at the refineries. A move to adopt catalyst vehicles is often largely dependent on the rate of renewal of the vehicle fleet, which can take a decade or so. In the USA, where unleaded fuel costs more than leaded fuel, the cheaper fuel has been used by many consumers in catalyst-equipped cars. The lead rapidly poisons the catalyst and it no longer works properly.

Photochemical smog
Vehicles, together with some industrial processes, are also responsible for the emission of hydrocarbons. As with carbon monoxide their concentration will follow the pattern of human activities, especially the use of motor

Table 7.1. *The enhacement of heavy metals in dust in London, where the enrichment factors are calculated as the ratio of the concentration of the element found in the London dust relative to its mean crustal abundance (from data of Schwar and Ball (1983))*

Element	Mean concentration (ppm)	Enrichment factor
Iron	41 400	0.828
Lead	800	62
Zinc	660	9.4
Nickel	570	7.6
Copper	140	2.5
Cadmium	3.4	17

vehicles. The reactions that take place within photochemical smog have been discussed in Section 6.5. Table 7.2 lists typical concentrations found in polluted atmospheres. Photochemical pollution was first observed in the USA and later in the sunny climates found in Australia and the Republic of South Africa, for example. In European cities these photochemical smogs were historically much less of a problem than smoke-laden winter fogs, but photochemical activity has increasingly been experienced during dry summer months. In Fig. 6.10 the diurnal variation of pollutants in photochemical smog is shown. In some cities, such as Athens, it has taken on a particular character that has earned it a local name, 'nephos', but summer haze has become a frequent experience across much of Europe.

Modern winter smogs
As has been mentioned several times already, the old winter smogs of sulphur dioxide and smoke have now almost vanished, at least from the cities of Europe, where control measures have been implemented.

Table 7.2. *The composition of a photochemical smog*

	Concentration (ppb)	
	Morning values	Other times
Carbon monoxide	10 000	
Nitric oxide	100	Mid-day low 20
Nitrogen dioxide	70	Mid-morning values 100
Dinitrogen trioxide	0.1	
Dinitrogen pentoxide	0.2	
Nitrous acid	1	
Ozone	50	Afternoon 200
Methane	2 000	
Ethane	50	
Ethene	30	
Acetylene	20	
Benzene	8	
Formaldehyde	50	
Other aldehydes	30	Afternoon 100
PAN	10–20	
	Concentration (molecule cm^{-3})	
Hydroperoxyl radical	10^{10}	
Hydroxyl radical	10^{8}	
Atomic oxygen	10^{5}	
Atomic hydrogen	<1	

However, there is evidence of a new type of winter smog. Nitrogen oxide emissions have risen considerably in line with increasing urban traffic. Now when air stagnates over European cities, concentrations of NO can become very high, many hundreds of ppb. If we couple high concentrations with low temperatures then the normally unimportant reaction

$$2NO + O_2 \rightarrow 2NO_2 \qquad\qquad (R7.1)$$

can be a significant source of NO_2. This is partly because the rate constant for this reaction increases as the temperature falls. As noted in Section 3.1, the reaction is second order in NO, so the rate increases dramatically with concentration. This is seen in Fig 7.4. At low concentrations the production of NO_2 becomes limited, by the amount of ozone present, at around 50 ppb NO_x. However, once the NO_x concentration has reached several hundred ppb, a square-law dependence on concentration is evident. This shows the importance of reaction (R7.1). The sensitivity to reactant concentration suggests that even small improvements in emission of NO when modern winter smogs are likely would yield great improvements in the NO_2 levels. Thus it might well be worth issuing alerts and attempting to reduce traffic and other urban sources.

A parallel contemporary problem is the increasing presence of nitrous acid in urban atmospheres. The mechanism of production is not well understood, although a homogeneous process is known:

Fig. 7.4. A plot of hourly mean NO_2 concentration as a function of the total NO_x concentration, London, July 1991 to July 1992 (Derwent and McMullen, 1993).

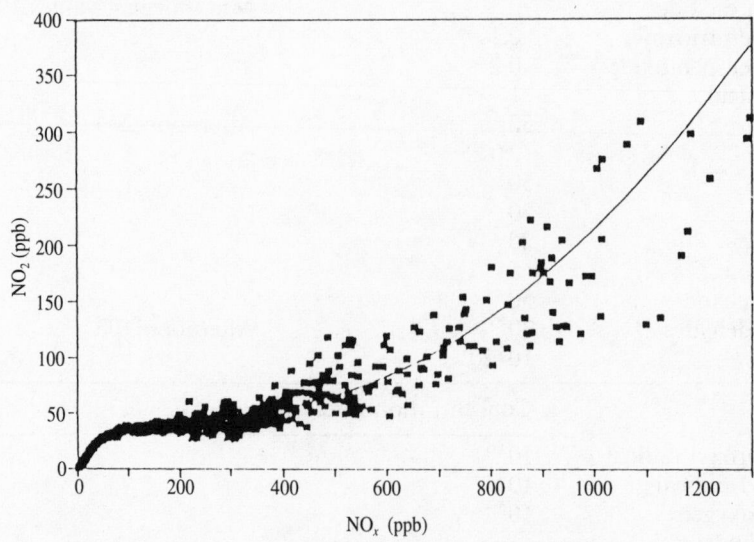

$$OH + NO + M \rightarrow HONO + M \tag{R7.2}$$

but the rate is slow. There is evidence that the reaction may be catalysed by surfaces. The production of nitrous acid could be important as it can act as an early morning source of OH through its photolysis:

$$HONO \overset{h\nu}{\rightarrow} OH + NO \tag{R7.3}$$

7.2   Health

The effect of air pollutants on health has been a major concern since the earliest times. We know from the Hippocratic Writings that Greek physicians were concerned with the role that air, water and locale played in human health. Roman doctors advised their wealthier patients to leave the city during bouts of illness and similar advice has persisted almost to the present day, although rural ozone and agrochemicals might well be changing this. It has been the effect of air pollution on human health that has concerned the civic governors of London from medieval times onwards. The early recognition of this link was probably related to contemporary beliefs that disease was carried by 'bad-airs'. The concerns led to the passage of the legislation against coal-burning as early as the thirteenth century.

Epidemiological evidence

By the seventeenth century, scientists had speculated on the nature of the agents responsible for the high death rates observed in the earliest demographical analyses of urban population. Coal smoke was thought to contain harmful arsenical or mercurial fractions. Doctors observed that these toxic elements had harmful effects on the health of populations living near metal mining and smelting regions. Other early scientists laid the blame upon the corrosive sulphur compounds present in coal and the high prevalence of lung disease in London as compared with Paris was of particular concern to those who knew both cities well. Wood tended to be a dominant fuel in Paris until the nineteenth century, so it was easy to presume that illnesses peculiar to London were the result of coal-burning.

London fogs have long been associated with increased death rates, although the relationship was not proved statistically until the present century. In the best known incident, the smog of December 1952, there were some 4000 excess deaths. However, re-analysis of early data shows that there were many similar occurrences in the nineteenth century. The meteorological data and public health statistics of the seventeenth century high death rates also show this association. The data in Fig. 7.5 show the deaths in London over several weeks in the winter of 1679. It can be seen that there were striking increases in deaths (among the elderly and from lung diseases) in the weeks that followed some 'great stinking fogs'. Of course there may be reasons other than air pollution, e.g. cold spells also kill elderly people. It is particularly striking that the second week of bad fogs, which occurred a little later in the year, did not result in a marked increase in deaths. It is probably because the susceptible population was eliminated in the first episode. This draws attention to an important feature of air pollution and health. It is often those people who are already in a poor state of health who are most susceptible.

No-one seems to dispute the fact that, in extreme episodes of the type discussed above, health may be affected by air pollutants, and people may even die. There are many incidents to prove this. However, there is a great deal of argument about whether the low levels of pollutants usually encountered in most cities are harmful. The relationship between air pollution and health is generally established based on statistical examination of the health records of urban populations. The main difficulty arises through the large number of variables involved and the dominant influence of smoking on lung diseases in surveys.

Fig. 7.5. The correlation between deaths in London and 'great stinking fogs' in the winter of 1679.

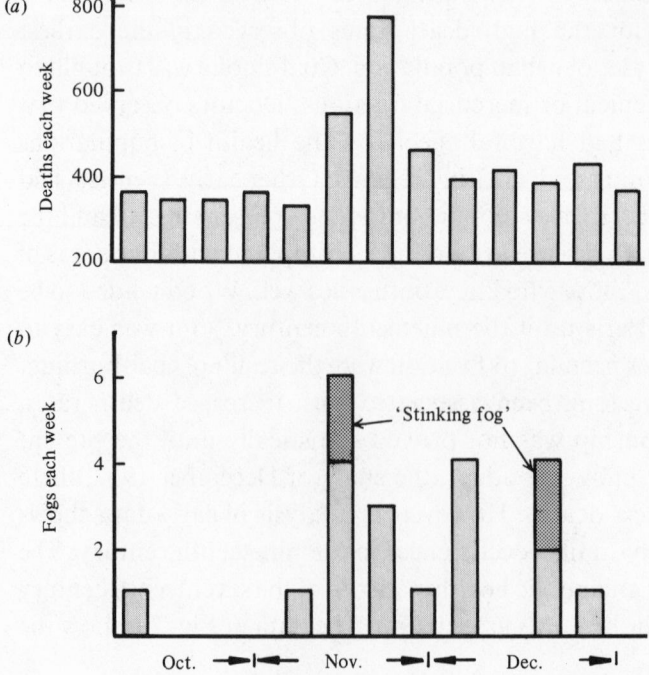

In recent years the evidence of the relationship between pollution found in urban air at low concentrations has improved. In Fig. 7.6 some important new evidence of the effect of various pollutants on the mortality rate is shown. The difficulty in undertaking these studies with urban populations is the range of confounding variables that are needed to consider issues such as location, occupational exposure, smoking habits etc. It is not simple to allow for such factors. However, when care is taken it is possible to examine the effect of various air pollution situations in different cities on an adjusted mortality rate. The data in Fig. 7.6 suggest a very strong linear correlation between concentration of fine particles and

Fig. 7.6. Adjusted mortality ratios for six cities in the USA as a function of different pollutant concentrations. P denotes Portage, Wisconsin; T Topeka, Kansas; W Watertown, Massachusetts; L St Louis, Missouri; H Harriman, Tennessee; and S Steubenville, Ohio. (Reprinted by permission of *The New England Journal of Medicine*, D. W. Dockery, **329**, 1753–9, 1993. Copyright 1993. *Massachusetts Medical Society*. All rights reserved.)

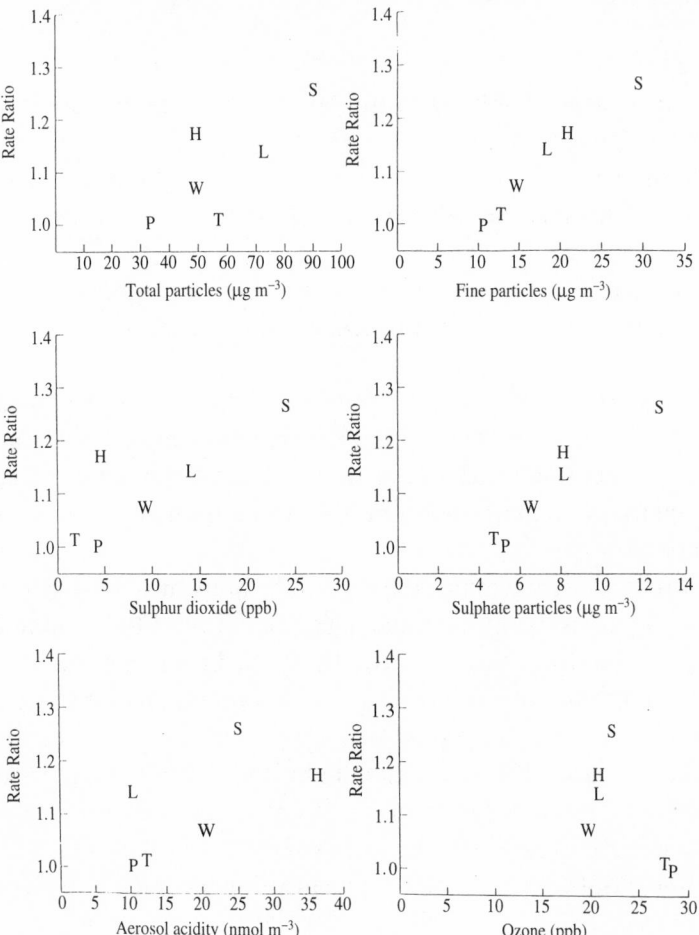

mortality. It may well be that minute particles mimic viruses and this increases physiological pressure and the likelihood of illness or death.

Although it is difficult to confirm the effects of low levels of air pollution on human health, it does appear that extremely low doses may be important. Carcinogens are often treated as if they have no threshold level. This means that there is no concentration, however small, below which there are no harmful effects. It has been argued that it is pointless to question the effect of air pollution on human health. The fact that we suspect it to be harmful should be a strong enough reason to undertake ameliorative measures. However, detailed knowledge is required for the allocation of resources. In a world of limited resources it is important to be able to answer questions such as that of which air pollutants are causing the most harm, so that the largest amount of attention can be paid to those. Even here the arguments are complicated by variations in both human physiological response to the pollutants and personal perceptions of what represents the most serious public health issues.

Sensitive areas

So far we have mentioned only respiratory diseases, but air pollutants also affect other parts of the body.

The skin appears comparatively untroubled at normal pollutant concentrations. This may be aided by the fact that the skin is frequently covered.

The eyes are sensitive to some air pollutants, but tear production helps by washing out irritants such as PAN. Smoke can exacerbate conjunctival diseases.

Hair offers a large area for absorption of pollutants and is usually exposed to the atmosphere. Although there are a few suggestions that air pollutants can affect the health of our hair, the most common problems are colour changes to hair dye and the more frequent shampooing required in cities.

The respiratory system remains the part of the body most susceptible to attack. Sensitive areas range from the nasal membranes to the lung (see Fig. 7.7). The respiratory system is broadly divided into an upper system (nasal cavity, pharynx and trachea) and a lower system (bronchi and lungs). The nasal cavity and the trachea are lined with mucous membrane that is ciliated in places. Fine hairs or cilia beat and move the superficial mucous fluid. Water-soluble gases such as sulphur dioxide can dissolve in the mucous, so that, at low concentrations (about 0.1 ppm) 50% can be removed in the nasal cavity. At high concentration as much as 95% may be removed. Some of this mucous, especially that in the trachea, will find its way into the gut. In the lower lung the bronchial passages become successively smaller through branching and terminate in alveoli. There are some 700 million alveoli, which, despite their small individual size, provide the lung with a surface area of more than 100 m² for the exchange of gases. The gases pass through the walls of the alveoli and into the blood.

Particulate material behaves a little differently in the respiratory system. The size of particles exerts an important control on the probability

Fig. 7.7. A schematic diagram of the human respiratory system. The deeper parts of the system are unciliated, so the residence time of particles in the alveoli may be several years.

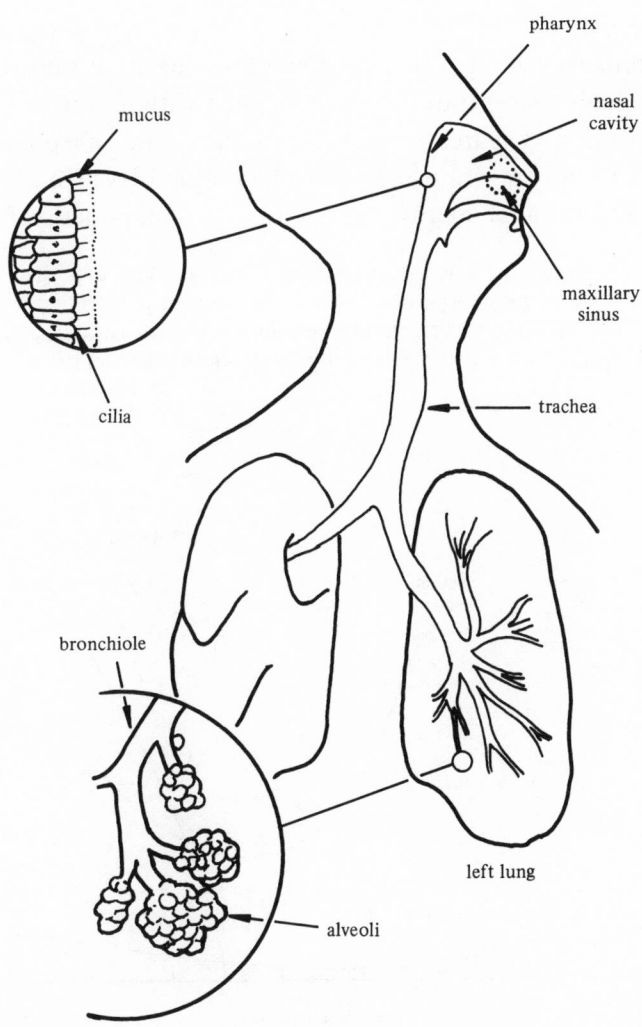

of transfer into the lung. In Fig. 7.8 where various particle size fractions are deposited within the respiratory tract is shown. Only the smallest penetrate into the lung. Some dissolution of solid material can occur so that it can enter the blood. This is of concern where the small particles, most often those from diesel engines contain toxic metals or organic substances.

Having considered the general effects of air pollutants, we will go through the various types of pollutant and examine them in more detail. However, because health effects of air pollutants are related to dose, many recommendations have to express both concentrations and length of exposure.

Carbon monoxide
This pollutant is present where carbonaceous fuels are burned. Often particularly high concentrations are associated with motor vehicles and interior pollution. Cigarette smoke contains some 400 ppm carbon monoxide, so smokers will usually have high blood levels of the gas. In addition, inhalation of methylene chloride (a component of varnish

Fig. 7.8. Fractional amounts of particles of various sizes deposited in the various areas of the respiratory tract. Note the high probability of deposition of micrometre-sized particles in the 'deep lung' (the alveolar area). This explains the special concern about pollutant particles in this size range.

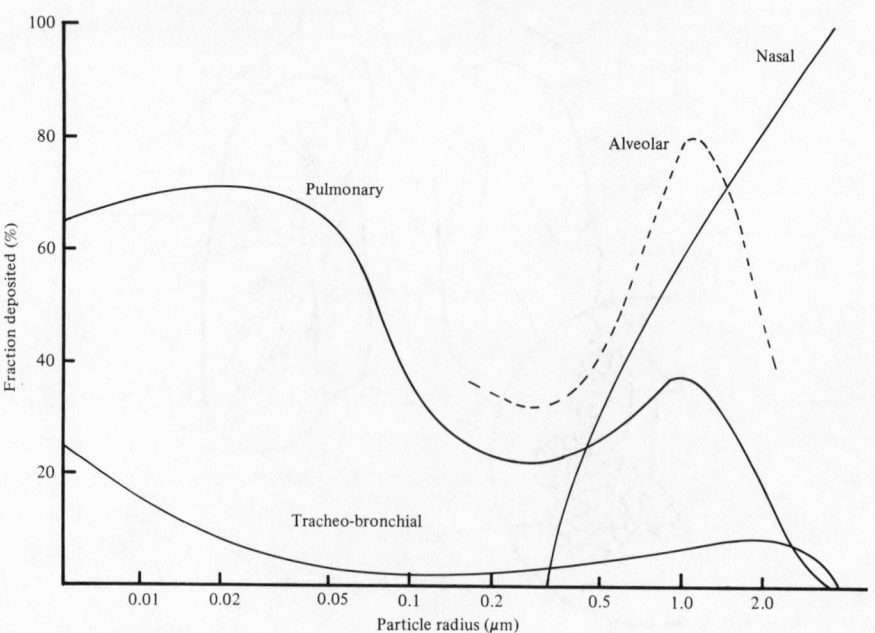

remover) can lead to high carbon monoxide doses as it is a product of the metabolism of methylene chloride. Carbon monoxide is particularly dangerous because it is odourless. The gas binds strongly with haemoglobin and lowers the oxygen-carrying capacity of the blood, by forming carboxyhaemoglobin (COHb). The amount of COHb in the blood is dependent on the concentration of the gas, level of bodily activity and the exposure time. The COHb concentration in blood is a convenient measure of the effect of exposure to carbon monoxide.

Carboxyhaemoglobin in the blood lowers vigilance and the capacity to perform work when its concentrations in blood exceed 5% of the total haemoglobin. It seems probable that angina symptoms may be aggravated in chronic sufferers of the disease at COHb concentrations in the range 2.9–4.5%. At higher concentrations 10–15%, symptoms such as headaches or dizziness are reported. The level of COHb recommended to protect non-smokers is 2.5–3%. Although converting this into exposure recommendations is sensitive to the level of activity, the World Health Organisation (WHO) suggests the following guideline values:

> 50 ppm for 30 min
> 25 ppm for 1 h
> 10 ppm for 8 h

Nitrogen dioxide
Nitrogen dioxide is an oxidising agent and associated with photochemical pollution. It has an olfactory threshold as low as 0.1 ppm. High concentrations, in the 1 ppm range, cause a whole range of reversible and irreversible effects arising from structural changes in cells of the respiratory system. There is some indication that, at 0.5 ppm, the risk of bacterial infection of the lung increases. Confusion often arises over the relative sensitivity of normal individuals, bronchial patients and asthma sufferers to nitrogen dioxide. Epidemiological studies of exposure to the NO_2 released during cooking with gas indicate an increased frequency of respiratory illness among children.

The recommended exposure to NO_2 has been difficult to set, but the small (10%) and statistically significant change in the pulmonary function of asthmatics exposed to 0.3 ppm has been used with other evidence to set the following guidelines:

> 0.21 ppm for 1 h
> 0.08 ppm for 24 h.

Sulphur dioxide and particles

The toxicology of sulphur dioxide seems surprising at first because the gas is not particularly harmful. Laboratory animals have lived quite happily for long periods at concentrations of 25 ppm. The concentration limit for 8 h industrial exposure has traditionally been 5 ppm. Yet, in classic air pollution episodes, deaths have occurred at concentrations of about 1 ppm.

This arises because sulphur dioxide is not especially toxic in isolation, but in the urban air it is present together with other trace components that amplify its toxic effect. Where the toxicity of two compounds increases because they are present together, they are said to act synergistically. The sulphur dioxide–smoke system is a complex one that can be broken into different elements. These are often considered to be

   (i) sulphur dioxide gas
  (ii) sulphuric acid aerosols, produced from the oxidation of sulphur dioxide
 (iii) sulphur dioxide and smoke particles.

Only small amounts of sulphur dioxide gas reach the lower respiratory tract, because it is so effectively deposited in the mucous membranes that line the nose and upper respiratory tract. The acid aerosols are associated with much smaller particles (0.3–$0.6\,\mu m$), and although they are likely to grow at the high humidities encountered within the respiratory pathways, they can be carried into the deeper lung airways. Some of this deposited acidity can be neutralised by the mucous, but this buffering capability is much reduced in the acidic saturated mucous typically found in asthmatics. Large fog droplets (10–$15\,\mu m$), by contrast, are deposited in the upper respiratory tract. A similar size–deposition dependence is expected from the soot particles that are usually associated with sulphur dioxide. However, with the declining concentrations of smoke and sulphur dioxide in many modern cities, future regulations may separate smoke and sulphur dioxide rather than treat their action synergistically.

Long exposure to low concentrations of sulphur dioxide leads to increased incidence of bronchitis, especially among smokers. Asthmatics show demonstrable health effects when they are exercising at levels as low as $1000\,\mu g\,m^{-3}$. Such populations are also sensitive to concentrations of acid aerosol as low as $100\,\mu g\,m^{-3}$. Acid aerosol and fine particles, known as PM10 (i.e. less than 10 μm in diameter), have become a recent area for regulation. It appears that fine particle concentrations are strongly correlated with mortality and a range of illnesses. Concentrations as low as $10\,\mu mg\,m^{-3}$ still show a statistical effect in large urban populations. However, such low concentrations will be difficult to achieve in most modern cities.

Experience gained from major incidents, such as the London smog of 1952, mean that the guidelines for sulphur dioxide and particulate materials have been established for some time. The WHO recommends that short term exposures of 24 h should not exceed 125 $\mu g\,m^{-3}$ for both sulphur dioxide and black smoke. In addition, the annual long-term exposure should be less than 50 $\mu g\,m^{-3}$ for sulphur dioxide and black smoke.

Ozone

Ozone in the air can be responsible for a range of respiratory effects in humans. Typically coughs, throat dryness, thoracic pain and increased mucous production are reported as a result of exposure. The results often show that healthy individuals are just as susceptible as asthmatics or cigarette smokers and that there is no threshold for an observed ozone effect. A decrement in pulmonary function in children may occur at concentrations as low as 220 $\mu g\,m^{-3}$ (0.11 ppm).

One difficulty in setting an appropriate guideline for ozone is the high background levels. After considering these the WHO recommended 150–200 $\mu g\,m^{-3}$ (0.076–0.1 ppm) in the 1 h range and an 8 h value 100–120 $\mu g\,m^{-3}$ (0.05–0.06 ppm). More recently, an air quality standard of 50 ppb over 8 h has been recommended for the UK. However, one worry is that short-term peak concentrations could have a greater health impact than 8 h averages.

There is some evidence of a synergism between ozone and acid sulphate particles. There are many arguments that the upward trend in asthma is related to atmospheric ozone, but studies have largely failed to reveal a consistent positive association between these two variables.

Peroxyacetylnitrate (PAN)

The peroxyalkylnitrates and peroxyarylnitrates are severe irritants for the eye. Although their role in photochemical smog has long been known, the WHO has not set guidelines. There are some suspicions that PAN may be carcinogenic, so recommendations about exposure are difficult. It is much less toxic than ozone with respect to its effect on the respiratory system, so the most relevant health effects remain irritation of eyes.

Benzene

Benzene belongs to a range of volatile organic compounds that are present in the urban atmosphere. Some of these arise from evaporative automobile emissions, but the use of solvents is also responsible for enhanced concentrations, especially in the indoor environment. Benzene is particularly associated with leukaemia. As it is a carcinogen, guidelines are set to consider long term average values. These have often been expressed in terms of the probability of contracting cancer from a cumulative exposure. The WHO estimated the lifetime risk of cancer from benzene to be 4×10^{-6} from benzene at an average concentration $1\,\mu g\,m^{-3}$ benzene. There are newer recommendations in the UK that aim to set the long-term guideline value at 5 pbb, ultimately declining to 1 ppb.

Other substances are also of concern: formaldehyde and other aldehydes (products of photochemical smog), styrene (industrial emissions) and 1,3-butadiene (a carcinogen from synthetic rubber production and diesels). Less regulated air pollutants, of this type, will become the focus of increased attention over the next few decades.

Metals

Some metals in particles can catalyse the oxidation of sulphur dioxide and thus enhance the toxicity of the sulphur(IV), by converting it to more harmful sulphuric acid. In this way there is a synergism between metal containing particles and sulphur dioxide. Metals in atmospheric aerosols also have toxic effects of their own. The smallest particles are often highly enriched in volatile toxic elements such as thallium, cadmium and lead, so these find their way deep into the lung. Particles emitted from non-catalyst automobiles will contain lead derived from the anti-knock agent in petrol. The deposition of these lead particles in the lung causes high blood lead levels. Lead has a range of health effects and there has been particular concern that elevated concentrations may affect the intelligence of urban children. The WHO has recommended 0.5–$1.0\,\mu g\,m^{-3}$ lead as a guideline with a two-fold safety factor.

Other metals such as nickel, chromium and cadmium are carcinogenic. Cancer of the lung and larynx is prevalent in workers in nickel refineries and observations of this kind have allowed the WHO to set a lifetime cancer risk of 4×10^{-4} from exposure to $1\,\mu g\,(Ni)\,m^{-3}$. There is plentiful evidence of increased incidence of lung cancer from exposure to chromates and a little evidence of similar cancers among cadmium workers.

Organic particulate material

Other particulate material will consist of a wide range of organic compounds. Polycyclic organic compounds (see Fig 4.10(*b*) for structures) are the most studied of these because so many are well known carcinogens, which distribute themselves between the gaseous and particulate phase in a manner dependent on their volatility (see Section 4.4 and Table 7.3). Of these benzo(a)pyrene is frequently studied as it also occurs in relatively high concentrations. There has been a decrease in the levels of benzo(a)pyrene in urban air over the last fifty years. This parallels the decrease in home fires as a source of heat since these emit particularly high levels of polycyclic aromatic hydrocarbons. Some increase in levels is to be expected in areas where the use of wood-burning stoves is rising. There is some concern about the possible health risks imposed by their increasing popularity. This is especially true of those that burn at a low temperature. Chlorinated and nitrated PAHs are now detected in the urban atmosphere. These substituted compounds often appear to be, biologically, more active than their parent PAH and no doubt include many particularly potent carcinogens.

A general summary of the WHO air quality guidelines and other regulations is given in Table 7.4.

Table 7.3. *Polycyclic organic compounds* ($ng\,m^{-3}$) *in urban air during* 1992 (*from QUARG* (1993))

Compound	London		Birmingham	
	Particulate	Vapour	Particulate	Vapour
Napthalene			0.21	13.0
Acenapthylene		0.61	14.8	
Fluorene			1.1	12.6
Acenapthene			1.6	11.9
Phenanthrene	0.11	5.01	1.1	23.0
Anthracene	0.18	2.66	0.4	4.1
Fluoranthene	0.81	2.65	1.2	11.2
Pyrene	0.79	3.00	2.4	35.6
Benzo(c)phenananthrene	0.85	1.98		
Cyclo(c,d)pyrene	2.61	1.47		
Benzo(a)anthracene	0.79	0.62	1.5	4.1
Chrysene	1.22	0.40	2.2	4.3
Benzo(b)fluoranthene	0.98	0.24		
Benzo(e)pyrene	2.00	0.31		
Benzo(k)fluoranthene	1.61	0.17	2.0	0.19
Benzo(a)pyrene	0.68	0.07	1.1	0.08
Dibenzo(a,h)anthracene	0.13		0.79	0.04
Benzo(g,h,i)perylene	3.30	0.01	1.9	0.06
Indeno(1,2,3,c,d)pyrene	1.57		2.0	
Anthrathene	0.63			
Coronene	1.67		1.0	

7.3   Effects on plants and animals

Although damage to human health has always been the most important implication of air pollution, worry about the impact on vegetation and livestock also has a long history. Some earliest modern legislation regarding air pollution aimed to lessen agricultural damage from the alkali industry last century. The hydrochloric acid that it produced devastated the vegetation over vast tracts of English countryside. London fogs of the late nineteenth century simultaneously caused many deaths among cattle brought to the metropolitan markets. During periods of fogs their suffering was so great that they had to be slaughtered. Not so much, an observer noted, 'to save their lives, but for the value of their meat'. It

Table 7.4. *Air quality guidelines suggested by the World Health Organization (WHO, 1987)*

Substance	Time-weighted average or unit risk[a]	Averaging time
Arsenic	4×10^{-3} (a)	
Benzene	4×10^{-6} (a)	
Benzo(a)pyrene	9×10^{-2} (a)	
Cadmium	$1-5\,\mathrm{ng\,m^{-3}}$	1a (rural)
	$10-20\,\mathrm{ng\,m^{-3}}$	1a (urban)
Carbon monoxide	$100\,\mathrm{mg\,m^{-3}}$	15 min
	$60\,\mathrm{mg\,m^{-3}}$	30 min
	$30\,\mathrm{mg\,m^{-3}}$	60 min
	$10\,\mathrm{mg\,m^{-3}}$	8 h
Formaldehyde	$100\,\mathrm{\mu g\,m^{-3}}$	30 min
Hydrogen sulphide	$150\,\mathrm{\mu g\,m^{-3}}$	24 h
Lead	$0.5-1\,\mathrm{\mu g\,m^{-3}}$	1a
Mercury	$1\,\mathrm{\mu g\,m^{-3}}$	1a (indoors)
Nickel	4×10^{-4}	
Nitrogen dioxide	$400\,\mathrm{\mu g\,m^{-3}}$	1 h
	$150\,\mathrm{\mu g\,m^{-3}}$	24 h
Ozone	$150-200\,\mathrm{\mu g\,m^{-3}}$	1 h
	$100-120\,\mathrm{\mu g\,m^{-3}}$	24 h
Styrene	$800\,\mathrm{\mu g\,m^{-3}}$	24 h
Sulphur dioxide[b]	$500\,\mathrm{\mu g\,m^{-3}}$	10 min
	$350\,\mathrm{\mu g\,m^{-3}}$	1 h
Toluene	$8\,\mathrm{mg\,m^{-3}}$	24 h
Vanadium	$1\,\mathrm{\mu g\,m^{-3}}$	24 h
Vinyl chloride	1×10^{-6} (a)	

[a]As lifetime risk from exposure to $1\,\mathrm{\mu g\,m^{-3}}$.
[b]Normally, one needs to consider this in terms of its synergism with smoke as discussed in the text.

sometimes seems that the fate of cattle prompted more sympathy than the bronchial complaints suffered by Victorian urban dwellers.

In North America damage to vegetation also sparked off considerable concern about air pollution. It is likely that the agitation that vegetation damage arouses is due to the direct economic loss that is suffered by farmers. The Consolidated Mining and Smelting Co. at Trail on the Columbia River in Canada was responsible for sulphur dioxide pollution that drifted across the border into the USA. Damage to crops began to be noticed in 1929, but it was not until 1941 that a report by an international commission of 1939 allowed a decision to be reached and the company ordered to pay US $350 000. Early this century, nickel refining began at Sudbury in Ontario. The process involved roasting the sulphide ore and produced enormous amounts of sulphur dioxide. Vegetation was destroyed downwind for a distance of 40 km.

Much of the initial research into the Los Angeles smog was prompted by the damage to leafy vegetables, especially cash crops such as lettuces. Initially, many thought that this was caused by sulphur dioxide, but the research established ozone in photochemical smog as a key causative agent. This allowed scientists of the early 1950 to determine the important role that petroleum vapour played in smog formation.

Vegetation is sensitive to air pollutants. Lichens, in particular, rapidly die out in areas where air pollution, especially sulphur dioxide, occurs. This becomes evident through a drop in diversity of the lichens growing in a given area. Some fourteen species of lichen in Epping Forest to the north east of London became extinct between the surveys made in the late eighteenth century and those a century later. Many more species have died out since, so that now only the most tolerant are to be found. This is because the concentration of sulphur dioxide in the forest have increased as London has expanded around it. Lichen diversity offers an excellent method of estimating the pollution concentration at a given site. It is possible to use this technique to estimate the mean sulphur dioxide concentration in Epping Forest over a period of two centuries (Table 7.5). The lichen diversity of England and Wales is shown in Fig. 7.9 and corresponds well with the distribution of emissions.

The relationship between damage to lichen diversity and average sulphur dioxide concentration is an over-simplification. In reality we would need to consider the number of fumigations of a given intensity and the duration peaks in exposure. In addition we need to consider that other air pollutants affect lichen, with fluorides showing a particularly marked effect. With the gradual reduction of sulphur dioxide concentrations in

Table 7.5. *Winter mean sulphur dioxide concentrations in Epping Forest as estimated from lichen diversity by Hawksworth and Rose (1970)*

Survey date	Sulphur dioxide (ppb)
1784–96	<10
1865–68	15
1881–82	20
1909–19	25
1969–70	35

Fig. 7.9. A map of the extent of pollution by sulphur dioxide compiled by studying the diversities of communities of epiphytic lichen. The more sensitive lichens are absent from polluted regions. Darker shadings indicate low diversity. (The map is simplified from Hawksworth and Rose (1970).)

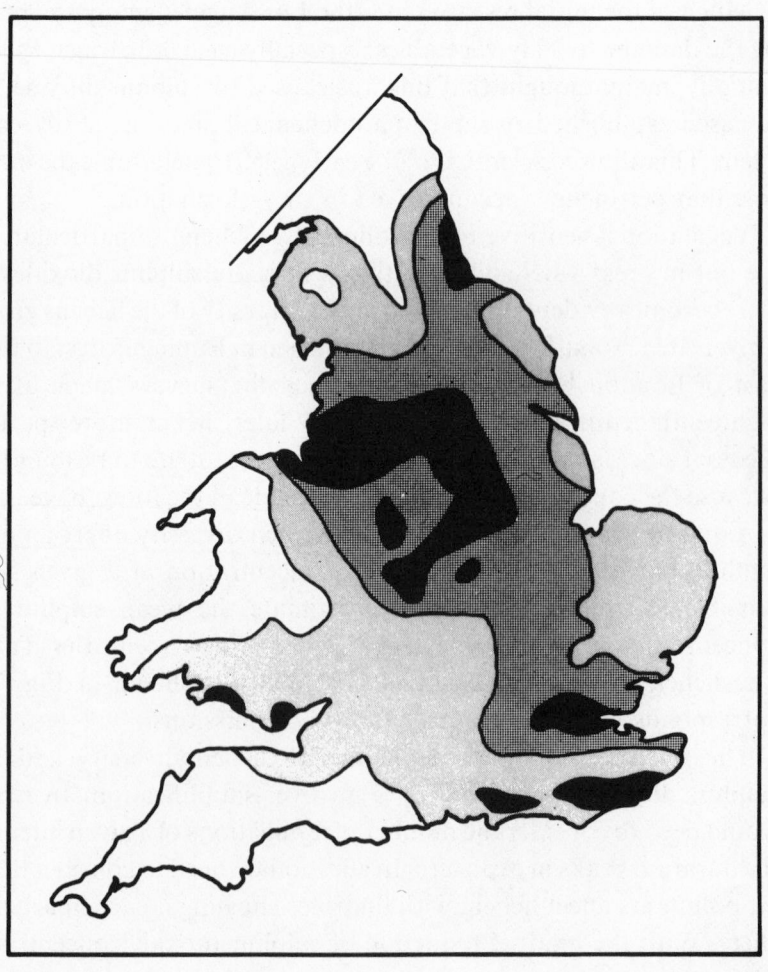

the air of European cities there has been a re-invasion of lichens into these areas. The air is still acidic, so it is acidophilic lichens that are most frequently found. Ozone has also been recognised as an important factor in the decline of lichen species in the San Bernadino Mountains, not far from Los Angeles.

Much interest has been shown in the effect of air pollutants on vascular plants, as these are of greater commercial importance than lichens. Air pollutants change the water balance of plants, which shows itself in terms of structural changes. High sulphur dioxide or fluoride concentrations, for instance, cause collapse and death of cells throughout the leaf. Rapid breakdown of chlorophyll can be evident as necrotic markings on the leaves. These arise because of the death of some tissue. The areas of damage often start as somewhat bruised or apparently water-soaked patches, which dry out to give a pattern that is often characteristic of the plant species or of attack by a particular pollutant (Table 7.6). Several other forms of damage can be seen. The loss of the green pigmentation due to chlorophyll (chlorosis) is especially common after exposure to sulphur dioxide and the leaves may even drop off. Alfalfa proves to be a most sensitive crop and exposure to sulphur dioxide can be evident as necrotic markings after just a few hours exposure at 300–500 ppb.

Ozone is the primary damaging agent of photochemical smog. White or tan flecks can appear on the sensitive plants, such as tobacco, after more than 3 h exposure at 200 ppb. The production of the radicals, HO_2^-, HO^+, O^- and OH_2^- in the water associated with the leaf results in the oxidation of various cellular metabolites. Ozone can impair photosynthesis at concentrations lower than that at which other effects, such as a loss of yield, are evident.

The reduction in crop yield is important because it may be occurring even where no obvious damage has been found. Exposure of corn to as little as 50–100 ppb ozone, for 6 h per day, through the growing season can cause reduction in yield. Such concentrations can be exceeded for

Table 7.6. *Necrotic markings on plants exposed to air pollutants*

Gas	Markings
Sulphur dioxide	Light coloured blotches between veins
Sulphuric acid	Pock marks on upper leaf surfaces
PAN	Metallic sheen to under surfaces of leaves
Ozone	Stippling to upper surfaces of leaves
Fluoride	Brown or tan colour at leaf tips

long periods of time during the summer in many continental areas. Brief periods of high concentration can be particularly critical for ozone damage to plants. Furthermore, as with human exposure to pollutants, an increasing number of synergisms are being found (i.e. SO_2 and O_3 act synergistically to increase the damage to crops). It may be that atmospheric pollutants also act by increasing plant stress and thence vulnerability to drought or disease.

So far we have concentrated on the effects of gases. Particulate material can also have severe effects on plant life. The soot that fell from the atmosphere in earlier times caused considerable damage to plant life. There are many observations of this in London over the centuries. When soot coatings are thick enough they can prevent photosynthesis and block the stomata of the plant leaf. Salt damage from wind-blown spray in coastal areas has been recognised as a problem by classical writers. Only more sensitive plants exhibit serious damage, but it is possible that crop yields are lower in such places.

Acid mists and sprays from chemical works can be serious locally, but they are, fortunately, not too difficult to control. The most serious problem appears to be the fluoride emissions from the combustion of high-fluoride coal, aluminium production or brick-making. Besides vegetation damage, cattle eating fluoride contaminated grass develop fluorisis, a bone disease. The reason that livestock are so sensitive is partially because grazing animals can eat vegetation that has such an enormous surface area for absorbing pollutants. In addition animals ingest considerable quantities of contaminated soil along their food.

Particulate materials also contain toxic metals and these may be found at high levels on the surfaces of plants near pollution sources. Detailed investigations of these particles have shown that they are transported to the vegetation canopy by wind eddies and then deposited by impaction. These pollutants can interfere with plant health, but often the real concern centres on the implications of such contamination for the consumption of roadside crops by humans or animals. Metal-containing particles are found in high concentrations on the vegetation growing in polluted locations. In Fig. 7.10 is shown the relative concentration of lead on blackberry leaves, as a function of distance from a roadway. With larger emissions, from smelters for example, toxic metal concentrations can be enhanced many kilometres away from the source.

In places where pollution levels have been moderately high for long periods of time, it is possible to find some species that have evolved tolerances to pollution. Rye grass populations in near-urban locations have developed a tolerance to sulphur dioxide; they are often found in areas subject to high SO_2 concentrations. Populations selected from rural areas are far less tolerant towards SO_2 than the urban grasses. Another case of pollution related evolution is found in the moth *Biston betularia*. This moth has several forms. A dark form was relatively uncommon before the industrial revolution. However, in many parts of England it is now common. This adaptation was necessary for the moth to be able to camouflage itself against besooted tree trunks in industrial Britain. With the coming of cleaner air, the light form is again becoming more common.

7.4 Effects on building materials

Air pollutants can attack a variety of inert materials. The most obvious and rapid effect is soiling, by deposition of soot. Complaints about this are known to have been lodged since Roman times. The change in colour of painted surfaces is proportional to both the pollutant concentration and the exposure time. However, the soiling varies according to the colour of the deposited material. Although abandoning coal as a fuel might be assumed to produce lighter coloured soot deposit, this is now counteracted by dark smoke from an increasing diesel fleet. Diesel particles are very

Fig. 7.10. The relative amount of lead deposited on blackberry leaves as a function of distance from the centre of a road. (The analyses were performed by S. Mears of the University of East Anglia.)

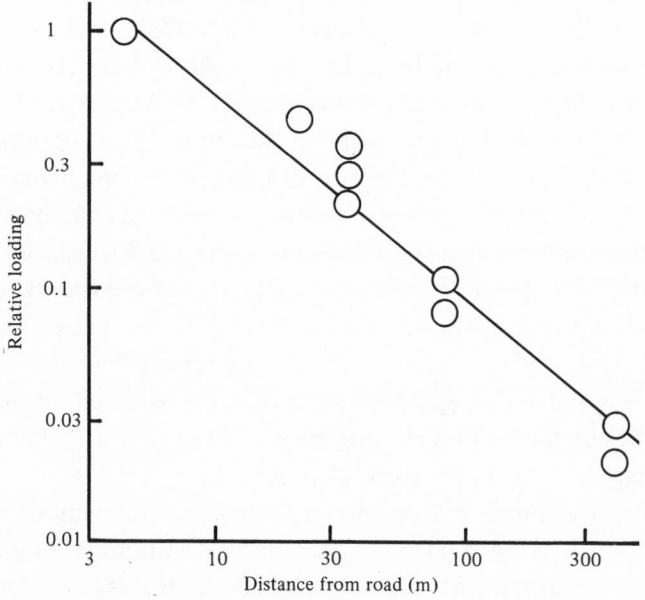

black and stick to surfaces effectively, re-soiling cleaned London buildings over a period of eight years. The degree of adhesion varies with both the chemical and the physical nature of the surface. Rough surfaces collect soot more readily than do smooth ones. Sandstones, for instance, accumulate soot very effectively and the rain will not wash it out. Consequently, the stone becomes uniformly blackened. On the other hand, limestones can be cleaned by the rain and the soot collects on sheltered surfaces.

Soot particles can also carry acidic substances with them and these enhance the corrosion of stone and metal building materials. This is a very general problem in polluted atmospheres. Traditionally, the most damaging acid in the urban atmosphere has been sulphuric acid derived from the oxidation of sulphur dioxide. One can imagine a simple reaction to account for weathering of limestone in an urban environment:

$$H_2SO_4 + CaCO_3 \rightarrow CaSO_4 + CO_2 + H_2O \qquad \text{(R7.4)}$$

This transformation of limestone to gypsum is serious because gypsum is more soluble than limestone and will dissolve more readily. It also has a larger crystalline volume, so the mechanical stress that results from the transformation can damage the building stone. Gypsum crusts are well known in association with stone buildings, so the importance of sulphur deposition has always been clear. Much of the most worrisome damage has accrued to older structures of great aesthetic and cultural value.

In recent years damage to the Acropolis has received much attention. The polluted atmosphere or 'nephos' of Athens is very different from the coal-burning atmosphere traditionally found in European cities. It is a product of the great rise in population and industry of Athens, and a parallel increase in reliance on the automobile. The atmosphere contains sulphur dioxide, but also it is a photochemical situation in which nitrogen dioxide and ozone are produced. Sulphur dioxide has long been recognised as responsible for the sulphate crusts that disfigure stone, but a role for the gases of the modern photochemical smog is not yet clear. One has only to note the extremely high solubility of calcium nitrate to recognise that atmospheres containing nitric acid can also be extremely harmful to building stone. It is possible that the nitrogen oxides could speed up the oxidation of sulphites on the building to sulphuric acids or that they make the surface nitrogen-rich and liable to biodegradation.

Current photochemically polluted atmospheres will attack more modern materials. Some sealants, nylon and rubber are damaged by photoxidants. Rubber is a polymeric material with many double bonds, so it is degraded and cracked by ozone. Tyres and windscreen wiper blades have been especially vulnerable to oxidants. Manufacturers have designed anti-ozonants and synthetic rubbers, for example, have double bonds protected by other chemical groups that can make them more resistant to damage by ozone.

7.5 Indoor pollution

It is not at all surprising that the earliest pollution we know about occurred indoors. The smoke from fires in poorly ventilated huts has been a problem since the earliest times. This is not idle speculation. Almost all the mummified samples of ancient lung tissue available for study show evidence of anthracosis (a blackening due to exposure to soot- laden atmospheres). Anthracotic pigment is found in mummies from the remote Arctic to Egypt. There are no paleo-pathological samples of lung tissue for people living in mid-latitude regions, but there was a high incidence of sinusitis among Anglo-Saxons and perhaps among some plains Indians of North America. Sinusitis can be recognised in skeletal materials by a roughening of the bone at the base of the sinus (see Fig. 7.7). The disease may have been exacerbated by smoky interiors. In the Anglo-Saxon period this would correlate well with the fact that the Anglo-Saxons cooked indoors in huts that lacked a chimney. Throughout history the battle with interior smoke and poorly designed chimneys has continued and many smokeless stoves and chimney grates have been proposed. There has been a heightened interest in indoor air pollution over the last decade, because of the likely impact that it has on our health and well-being.

While many measurements of pollutant concentrations are available, relatively few of these pertain to the indoor environment where we spend so much of our lives. Contemporary interiors often have low ventilation rates to conserve energy, but increase our exposure to pollutants generated indoors. However as a general observation the concentrations of outdoor air pollutants are much lower indoors. There is little agreement on just how significant this decrease is, but it arises through absorption of the pollutants onto indoor surfaces. Sulphur dioxide and ozone appear to react with the surfaces available in rooms and decline in concentration rapidly if the room is closed off from the ambient atmosphere. Half lives of less than 1 h have been measured for SO_2 and as little as 10 min for O_3. The rates are a function of temperature, relative humidity and the area of reactive surfaces.

Air-conditioning can be effective in reducing the pollutant concentrations, especially for particulate materials. In sensitive interior environments, filters are often installed within the air-conditioning system. However, air-conditioning systems, especially where they are poorly maintained, can sometimes increase the concentrations of dusts and viable particles in the air.

Combustion processes, which often accompany cooking and heating, are readily identifiable as sources of air pollution. Gas cookers have a marked effect on the concentrations of NO_2 found in kitchens. A comparison between similar electric and gas kitchens has shown that the electric kitchens may have NO_2 concentrations some seven times lower (70 ppb as compared with 10 ppb). Wood, coal and other fuels can yield a particularly complex mixture of compounds during combustion. Measurements in kitchens in India, where wood, dung or coal was burned, have shown particularly high concentrations of the carcinogen benzo(a)pyrene. There could also be other air pollution hazards associated with kitchens, such as the carcinogenicity of airborne droplets of cooking oils.

Domestic heating causes many of the same types of problems as cooking. The conditions under which heating is required are also those under which it is desirable to keep ventilation to a minimum. Therefore, pollutants such as nitrogen dioxide, sulphur dioxide and carbon monoxide can easily reach high concentrations. Heating with wood stoves is also a matter of concern. The high-temperature burning found in a normal open fire effectively destroys the carcinogenic compounds produced by destructive distillation of the wood. However, in a slowly burning wood stove, with the minimum intake of oxygen, these compounds can be distilled up into the chimney or leak out into the room. Is the rather homely wood stove a far more dangerous polluter than heating by gas or electricity? Primitive dwellings do not escape the problems either, as measurements in Nepalese huts have shown that fuels such as yak dung, pine needles and wood can give measurable concentrations of lead and copper in the unventilated interior.

A more contemporary problem is the possibility of high formaldehyde concentrations within the interior of buildings. This occurs predominantly in modern structures fabricated from chip boards that are produced using formaldehyde-based resins, or less frequently from the urea–formaldehyde insulation foams that are used to fill wall cavities. In some measurements, indoor concentrations greater than 1 ppm have been found. This is of particular concern in the light of evidence that formaldehyde can be a carcinogen. However, there are more obvious short-term symptoms such as eye and respiratory complaints, together with headaches, nausea and sleeplessness. Cigarettes are a further source of formaldehyde indoors and produce significant amounts of carbon monoxide. This source has proved of particular concern within nuclear-powered submarines where smoking can push carbon monoxide to concentrations of 5–15 ppm.

The large range of solvents used in the manufacture of many indoor building and furnishing materials leads to high concentrations of volatile organic substances in the air of buildings. This is particularly true when the buildings are new. A hint of the range of materials that can be introduced into air in this way is evident from the list in Table 7.7. These types of organic compounds have been associated with 'sick building syndrome' (SBS), but it would be fair to say that SBS is not always easily attributed to chemical factors alone. There are suggestions that psychological issues relating to occupying and working in the building play a role in causing adverse human response to changes in indoor air.

The problems of internally generated air pollutants could easily be solved, it seems, by throwing open the windows and letting in more air. The Victorian approach to hygiene chemistry was one of trying to maximise light and air. Considerable effort was expended determining the maximum levels of carbon dioxide that should be allowed to accumulate in public rooms. Notions of 'lots of fresh air' always involved the assumption that lots of fresh air was available. There are times when it is reasonable to doubt that fresh air can be found outside!

There are also hazards associated with some recreational interiors. Lead can be generated from the use of small-bore rifles on indoor ranges. It is said that high concentrations of organic particulate materials are found in beer halls and at pop concerts. However, it is in the working environment that a particularly wide range of air pollutants is found. They include industrial solvents, many of which are suspected of being carcinogenic, and acid and metal vapours. In offices, solvents are in frequent use and photocopiers represent a source of ozone. The transfer of ozone into aircraft interiors from the outside air at high altitude has worried civil aviation authorities. Even the non-urban worker is subjected to a wide range of air pollutants: the pesticides in the air and ammonia and hydrogen sulphide that are regularly found in barns for animals. Ammonia can rise to concentrations in excess of 50 ppm when ventilation is low and the interiors heated. Such high levels can produce illness in the farm animals and effects on the farmworkers can hardly be absent.

The subject of industrial hygiene is complicated because of the enormous range of possible contaminants, so the interested reader should follow it up in more specialised texts. It is important not to limit concern to the most obvious manufacturing environments. For example, the air of

Table 7.7. *Indoor sources of VOCs from consumer materials*

PARTICLE BOARD
acetone, hexanal, propanol, butanone, benzaldehyde, benzene and simple carboxylic acids.

LATEX PAINT
2-propanol, butanone, ethylbenzene, propylbenzene, 1,1'-oxy-*bis*-butane, butylpropionate and toluene.

PAINT
acetaldehyde, acetone, propanal and butanone.

WOOD STAIN
2-butanone, benzene, trimethylbenzene, cyclodecane and branched and straight chain hydrocarbons.

CARPETS WITH STYRENE-BUTADIENE-RUBBER BACKING
4-phenylcyclohexene, 2,2,6,6-tetramethyl-4-methylideneheptane, styrene and 4-ethenylcyclohexene.

CARPETS WITH PVC BACKING
vinyl acetate, acetic acid, 2,2,4-trimethylpentane and 1,2-propandiol.

CARPETS WITH POLYURETHANE SECONDARY BACKING
hexamethylcyclotrisiloxane and the antioxidant 2,6-di-*tert*-butyl-4-methylphenol.

WOOLLEN CARPETS
carbonyl sulphide and carbon disulphide from wool protein.

FURNITURE POLISH
methylated pentanes, hexanes and heptanes, ethylbenzene and limonene.

ROOM FRESHENERS
nonane, decane, undecane, ethylheptane, limonene, dimethylchloride, 1,1,1,-trichloroethane and *p*-dichlorobenzene.

FLOOR WAX
nonane, decane, undecane, methlyoctane, methlynonane, 2-methlynonane, trimethylbenzene, *p*-ethyltoluene and trimethyl benzene.

POLYURETHANE FLOOR FINISH
nonane, decane, undecane, butanone, ethylbenzene and limonene.

FLOOR ADHESIVES
nonane, decane, undecane, dimethyloctane, 2-methlynonane and dimethylbenzene.

LATEX CAULK
methyl ethyl ketone, butyl propionate, 2-butoxyethanol, butanol, benzene and toluene.

ADHESIVES
alkane and cyclo-alkane compounds with a few simple aromatic compounds, e.g. toluene and styrene.

WALLPAPER GLUE
1,1,1-trichloroethane.

DRY CLEANED FABRICS
perchloroethylene.

STEEL WOOL SOAP PADS
ethylbenzene, xylenes and normal C-10 to C-12 alkanes.

DETERGENTS
ethanol, monoethanolamine and 1,2-propandiol.

beauty parlours is commonly contaminated by solvents used to apply and remove nail varnish and bakeries emit 10 g of ethanol for every kilogram of flour used, while carpentry shops are contaminated with wood dust.

Even within domestic settings there have been unusual sources of indoor air pollution. Green copper arsenite pigments (Scheele's Green and Paris Green) were used in interior paints and wall papers from the latter part of the eighteenth century. Under humid conditions moulds can grow on the walls and metabolise the arsenic within the pigments to trimethylarsine, a volatile and poisonous compound that may have been responsible for illnesses or even deaths. Historians have been concerned about whether Napoleon was poisoned with arsenic on St Helena. It is possible that the arsenic found to have accumulated in preserved samples of his hair had accumulated through volatilisation from the walls of rooms he occupied.

7.6   The museum environment

The previous section has discussed interior air pollution in general. This section examines a particular interior environment, where contaminants in the air can present very special problems. Museums and libraries not only hold large and important collections, but they also assume the responsibility for safely storing these objects for long periods of time. This is not an easy task when one considers that many objects are made of sensitive materials. Two environmental parameters that have traditionally concerned conservation scientists are temperature and light, both of which can damage sensitive artefacts. However these we will neglect, in order to concentrate on hazards of a more chemical nature.

The most troublesome contaminant is the seemingly benign atmospheric water vapour. Considerable effort is expended in galleries and storage facilities to maintain artefacts at a stable relative humidity. However, this does not always mean an absolutely dry environment, although metals are generally stored under fairly dry conditions. Iron objects are often kept at about 20% relative humidity and copper objects perhaps nearer 30%, to reduce corrosion. Wooden objects are stored at about 50–55% relative humidity and oriental organic objects sometimes higher at 60%.

The problem with water vapour is not really a function of its presence in air, so much as the rapid changes it undergoes through a typical day. As organic objects lose or gain water they are liable to warp, bend or stretch. Two common methods to control relative humidity in museums are (i) regulation of the gallery air through air-conditioning or (ii) storage of objects within cases with a stable hygrometric environment. The first is a very reasonable approach for objects such as paintings as they are best kept at close to normal room humidities and it is desirable to view them without a covering. Metal objects, on the other hand, would require a room that was much too dry for human occupancy, so they are often placed in cases where the humidity can be maintained a low values. The museum case has some further advantages. Obviously it provides some security besides a stable humidity, but it also protects the objects from other contaminants in the air.

Particulate material, or just plain dust, proves to be a contaminant of almost as much concern as water. Once again, air-conditioning and tightly sealed cases are good approaches to protection. To a lesser extent, the same could be said for protection against gaseous contaminants such as ozone or sulphur dioxide. However, the removal of air pollutants during air-conditioning is usually a costly undertaking. The museum case creates a small micro-environment, and thus seems to offer a more certain and less expensive route to protection against many air pollutants.

Cases are so widespread in museums that their interiors are now beginning to be studied as specialised indoor environments. The interior humidity of cases can be stabilised by the presence of large quantities of a humidity buffer such as silica gel. If conditions are such that the case interior dries, then the silica gel will release moisture to the air; the reverse will happen if the case interior becomes more humid. Large modern open-structured buildings where restaurant and working areas share a common air volume with the display galleries are likely to have contaminated air because of the wide range of activities taking place in the same air volume. This has made well-sealed cases a frequent method for protection against the air pollutants within galleries. Water vapour has a residence time of about a day, in a typical museum case. A gas like nitrogen dioxide, by virtue of its higher molecular weight, would diffuse in more slowly along the case seals, so the case buffers the interior against the more extreme fluctuations in external air pollutant concentrations. It is also important to remember that the concentration of outdoor air pollutant levels is generally lower inside buildings than in the outside ambient air anyway.

In this light, cases seem to provide an excellent solution to air pollution. However, a critical problem has so far been overlooked. If we seal an object in a case then it is in a very restricted volume of air. If even a small amount of pollutant were released inside the case, then the restricted volume would mean that it could achieve very high pollutant concentrations. Unfortunately, this can easily happen. Many materials used to construct museum displays have an alarming tendency to give off harmful vapours

Table 7.8. *Gases found in museums and museum cases and some of their probable effects*

Gas	Source	Effect
Formic acid	Wood, glues	Zinc and lead formates have been observed on metals, calcium and sodium formates on shells
Formaldehyde	Wood, glues	Oxidized to formic acid and behaves as it does
Acetic acid	Wood, cellulose acetate, PVA	Calcium and sodium acetate encrustations
Hydrochloric acid	PVC	Corrosion
Hydrogen sulphide	Dyes, glues, rubber	Tarnishes metals
Carbonyl sulphide	Wool, feathers	Tarnishes metals
Nitrogen oxides	Cellulose nitrate	Corrosion
Sulphur dioxide	External	Damages paper and leather
Ozone	External	Damages plastic and rubber

as they age, or the objects on display may themselves undergo changes that release gases. Some of these gases are corrosive; others tarnish metals. Air pollutants of particular concern to conservators are listed in Table 7.8 together with their effects.

Another very specialised interior is the library or archive, where sulphur dioxide can damage paper. The paper becomes brittle and yellow. It seems that, with sulphur dioxide, it is its oxidation product sulphuric acid that is the main cause of injury. Metals present in trace amounts in paper may be responsible for accelerating the damage. There are methods of treating paper to make it more resistant to damage from pollutants, although this is expensive. In important archives sulphur dioxide can be scrubbed from the incoming air. Sulphur dioxide also attacks the leather bindings of older books. Today, increasing attention is given to the careful choice of a range of acid-free papers and suitable bindings and repair materials for archive use.

Some modern industries, especially those manufacturing electronic devices and photographic materials, find that indoor air pollution can degrade products. Such sensitive processes require very clean conditions to be maintained by controlling air quality in critical areas of the factory.

7.7   Acid rain and acid deposition

Acid rain became a key environmental issue in the 1980s. Surprisingly, much was written about the problem in the 1960s and 1970s, but it attracted none of the interest that characterised the more recent decade. The precipitation monitoring network set up in Europe in the mid-1950s convinced many scientists that the pH of rainfall in Europe was on the decline. In theoretical terms, the dissolution of carbon dioxide in rainwater should yield a solution with a pH of about 5.6 (see Section 5.2). However, measurements made even in remote and unpolluted areas suggest that rain can exhibit widely ranging pH values. Alkaline rain has also been found. It arises from the presence of alkaline dusts in precipitation. Naturally occurring acidic precipitation can form through the presence of organic acids or the oxidation products of naturally emitted sulphur compounds. However, natural rainfall acidity hardly accounts for the pH values of less than 3.0 that are sometimes observed today.

Initially the term 'acid rain' was used to describe precipitation with anomalously low pH values. However, over the years writers also used it to describe the dry deposition of gaseous and particle pollutants. This is often called 'acid deposition', because it can be useful to distinguish between the general effects of deposited acidity that can occur quite close to pollution sources and the more distant effects of acid precipitation, whereby sulphuric acid formation can make acidification important more than a thousand kilometres from the source. Popular usage has sometimes broadened the term 'acid rain' to describe air pollution in general.

The precipitation pH of rainfall in the British Isles, Germany and the Low Countries currently lies close to 4.0. This suggests a marked decrease from the natural pH of rainfall. In Europe and in important regions of North America the emissions of sulphur dioxide are in decline, so a decline in the incidence of acid rain is to be expected. The decrease in deposition of sulphur in eastern Britain has been very pronounced. For the first time in decades, sulphur is being added to agricultural land as a nutrient. There are now many places in Europe and North America where decreases in sulphur deposition are observed, but these gains can easily be offset by increasing nitrogen oxide emissions.

Changes in the types of fuels used, economic shifts and desulphurisation have lead to reductions in sulphur emissions, but have often increased the importance of NO_x emissions, particularly those from vehicles (Fig. 7.11). Atmospheric acidity is increasingly derived from nitrogen rather than sulphur compounds, which has led to rising nitric acid deposition. This is often reflected by an increase in the HNO_3/H_2SO_4 ratio in the rainfall of industrialised nations.

The HNO_3/H_2SO_4 ratio is also influenced by distance from source. This is because oxidation of NO_x to nitric acid can proceed more rapidly than oxidation of SO_2 to sulphuric acid. Thus we could expect much higher ratios of HNO_3/H_2SO_4 in relatively young air masses. Hydrochloric acid can be produced as a primary pollutant from the combustion of high-chlorine coals. It has received little attention, but it is likely to be washed out relatively close to the source because it is very soluble. It is a particularly effective acid for attack on metal and paintwork which could make it the chemical explanation for such shocking headlines as 'BLACK RAIN EATS CARS'.

Fig. 7.11. (*a*) Global sulphur emissions from various fuels as a function of time (Ryaboshapko, 1985). (*b*) Per capita emissions of sulphur and nitrogen oxides in the USA (Galloway, 1989). The horizontal axis scale is the same for both graphs.

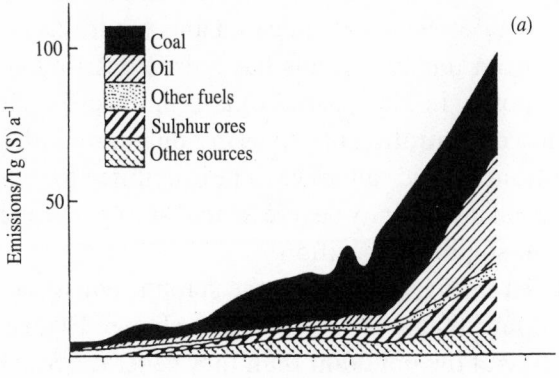

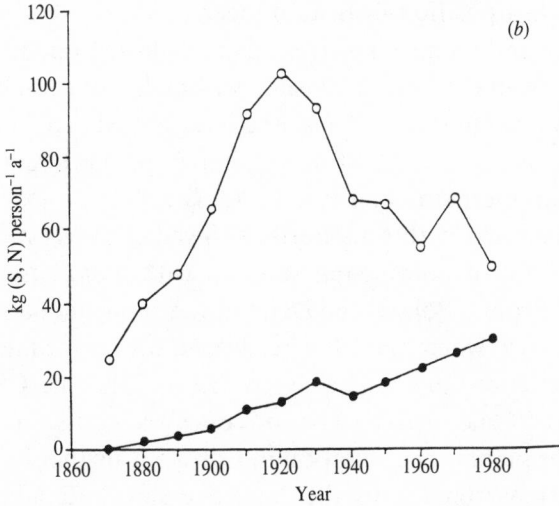

One important feature that emerges from analysis of precipitation data is the fact that high acidity often arises from a small number of rainfalls. The monthly precipitation acidity can be determined by just a few extremely acid samples. The popular term 'acidic events' reminds us that much of the strong acid is deposited at such occasions.

In its classic form, the acid rain problem can be attributed to increases in the emissions of SO_2 from industry and power generation. The long-range effects of acid rain may have been exacerbated by the 'tall stack' policy instigated to overcome local pollution problems. Section 6.7 illustrated how high chimneys lowered the concentration of pollutants at ground level. There are good reasons to believe that the pollutants from tall stacks are carried very long distances. However, the emissions have other features. For instance, the alkaline materials present within coal are all removed from the stack effluents, or at least fall out locally. Thus coal, which can be a neutral material, is separated into an alkaline fly ash, which is removed, and acid gases, which are released and can travel large distances. Alkaline dusts from unpaved roads have also contributed to increasing the pH of deposits in the past. Long-term records show evidence of a decline in the concentration of alkalis in rainfall, so both the acidic and basic contributions to acid rain have to be accounted for when predicting pH. Recent declines in alkalinity have lessened the improvements gained through declines in sulphur deposition.

The map of Europe given in Fig. 7.12 shows the sulphur emissions of various countries. The sulphur that they receive is also shown. Evidently some countries receive more of the pollutant than they generate. Scandinavia, with its high rainfall, receives a large amount of the pollutant sulphur. The isolated and windward position of Britain means that, while it is a very high emitter of sulphur dioxide, it receives relatively little from other countries. Some sulphur from the UK is deposited within the country, but a significant portion is carried to the northeast and is deposited in Scandinavia. Norway receives just over 20% of its sulphur from the UK and substantial amounts from Germany and Russia. Only 5% of Norway's deposited sulphur is generated within its borders. Sweden, being more easterly, is influenced more by continental sources, with major inputs from Germany, Sweden, Britain, Poland and Denmark. A similar situation has arisen in North America, where the USA has been a source for much of the acidification experienced in Canada. Both the UK and the USA have only slowly agreed to take active steps to reduce the emissions of sulphur from their power plants. The situation in Europe is further complicated by the high emissions from Russia, the Czech and Slovak Republics

Fig. 7.12. The emissions (E) and deposition (D) of sulphur for some countries in Europe (*c*. 1980). (Data are from the UNECE EMEP Programme, 1983).

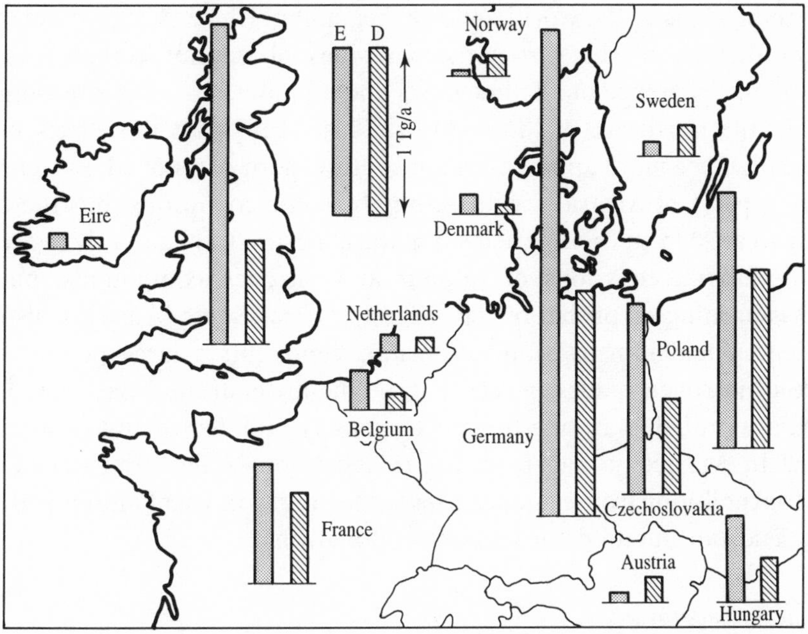

and Poland, where large amounts of high sulphur coal are burned in relatively unsophisticated plants. The difficult economic situation in these countries restricts funds available for lowering the emissions.

Acid rain has usually been attributed to industrial sulphur dioxide emissions and to a lesser extent to anthropogenic nitrogen oxide emissions, but other sources are possible. Frequent algal blooms that arise in the North Sea, perhaps as a function of its enhanced nutrient load, can lead to large fluxes of DMS to the atmosphere at some times of year. This DMS, after oxidation to sulphuric acid, could make a seasonal contribution to the acid burden at near coastal locations. Another reason for high atmospheric acidity could be that increasing amounts of ozone in the atmosphere could enhance the oxidation rate of sulphur dioxide through to sulphuric acid. In some regions large amounts of volatile organic compounds from human activities can be oxidised to organic acids and further add to the acid burden in precipitation.

China and the tropics

In the 1970s and 1980s interest in acid rain focussed on North America and Europe, but there is now evidence that rapid industrialisation in other parts of the world is giving rise to problems from acidification. As some of these regions lie in the tropics, the chemistry of acid rain may be different to that experienced in more temperate climates. As yet the programmes to study these areas of the world are only beginning. China is of particular interest because of the very large emissions of sulphur compounds, although the presence of alkaline dusts and ammonia has a neutralising effect in the northeast of the country. This means that the effects of acidification are more apparent in the southwestern parts of China. The northern parts of Australia witness highly acidic conditions, but these appear to arise from organic acids of natural origin. In Nigeria there are small industrial emissions of sulphur and nitrogen compounds, but biomass-burning is probably the largest source. South America also experiences acid rain as biomass-burning represents a large potential source of nitrogen oxides in Venezuela. The areas around São Paulo, Brazil, are probably among the most intensely polluted of any tropical region. In Mexico it has been argued that long-range transport of industrial pollution from major centres is affecting the ancient monuments in the Yucatan hundreds of kilometres downwind.

Damage to vegetation
The effect of pollutant gases on plants was dealt with in Section 7.3. Oxide emissions from combustion processes are known phytotoxins and have been responsible for extensive damage in the areas near smelters and large combustion sources. Both SO_2 and NO_x undergo extensive dilution with distance from the source and would be expected to become low in concentration. However, NO_x in particular is an important precursor in the generation of ozone and other oxidants. It is probably the catalytic role of NO_x that is important in damaging vegetation at large distances from its source. As we saw in the earlier section, ozone and other oxidants are phytotoxins, so these pollutants are likely to have significant effects on plants several hundred kilometres downwind. This may be particularly serious in the light of evidence that there could be synergistic relationships between the phytotoxicities of sulphur dioxide, nitrogen dioxide and ozone.

Besides these gaseous toxins, a polluted air mass will yield acidic precipitation. It has proved much harder to show the harmful effects from acid precipitation on vegetation than from the pollutant gases. Some would argue that direct damage to plants arises principally through dry deposition of the gaseous pollutants and that acid rain acts indirectly. For instance the acid precipitation may be responsible for acidification of the soil, especially in poorly buffered heathland soils.

Soil acidification can lead to changes in soil biota and the mobilisation of heavy metals. In particular, cadmium, which is not fixed within the soil structure, is likely to become mobilised by decreases in pH. However, it is important to realise that soils can be acidified through both natural processes and various types of human activity. Natural acidification can arise through the presence of the humic acids and carbon dioxide. Agriculture can increase the acidity considerably by using nitrogenous fertilisers. This might be represented by the equation

$$NH_4^+ + 2O_2 \rightarrow 2H^+ + NO_3^- + H_2O \qquad \text{(R7.4)}$$

Forestry can also acidify soils as trees remove nutrients such as calcium from the soil and replace these nutrients by H^+. If the whole tree is removed to be used as wood then this process represents addition of acid to the soil. Overall, it is likely that the increased use of fertilisers is the most serious agricultural source of protons. However, studies of large catchments suggest that the historical changes that we see in Europe are predominantly due to deposition of air pollutants.

Acid mists cause direct damage through attack on the leaf although the concentrations of acids have to be high – pH values less than 3.0 being required to lower the productivity of agricultural crops. It is possible that acid mists make the leaves more susceptible to damage from other air pollutants particularly if the acids are concentrated on the leaf surface by evaporation. The damage arising from deposition of acids onto leaves will usually appear as foliar lesions or leaf abscission.

The decline of forests in Europe and North America has probably been the most publicly recognised effect of acid deposition. In central Europe trees were progressively dying over some million hectares. In the worst hit areas in Bavaria, about 30–40% of the firs on south-facing, semi-arid slopes were found to be dying. Similar observations of acute visible damage continue to be made.

In the late 1970s acid rain, particularly in terms of its ability to leach base cations from the soil, or perhaps to mobilise toxic aluminium, was seen as the principal cause of forest damage. However, damage was also apparent on well buffered soils, so there was a shift to theories that it was ozone that bore much of the responsibility for forest decline. In some areas forest decline is clearly the result of foliar magnesium deficiency. However, the role of acid deposition in lowering the availability of magnesium is not particularly well understood.

As more work on forest decline reaches completion, it has been necessary to recognise that it is not really a single disease with a single cause. In one species, the Norway spruce, there are five recognisable types of damage: (i) needle-yellowing (in the Black Forest of Germany), (ii) crown thinning with little chlorosis (medium to high altitudes in central and northern Germany), (iii) needle reddening (older stands of the Bavarian Alps), (iv) chlorosis of needles (calcareous terrain above 1000 m in the Bavarian Alps), (v) crown thinning with reduced growth (in the northern German plain).

This complexity has led to several hypotheses for the causes of forest damage. Magnesium deficiency is thought to explain the needle-yellowing in the Black Forest. Multiple stress is often suggested as a cause of greater plant sensitivity to drought, nutrient deficiency and pathogens. Soil acidity and related aluminium toxicity continues to be seen as a possible cause. More recently there has been increased interest in the effects of enhanced nitrogen deposition, which is more than that required for plant growth. In the Netherlands this has been shown to increase susceptibility to stress. It is likely to be an explanation for the type of crown-thinning and reduced growth observed on the northern German plain. We must add to this the importance of ozone and acid mists that has already been discussed.

The summary in the previous paragraph examined only the most widely recognised theories of forest decline. There are other potential phytotoxins in rainfall that could affect plants across large areas of Europe and North America These have included (i) alkylated lead salts derived from automotive emissions, (ii) hydroxymethylhydroperoxides produced from oxidation of aqueous formaldehyde and (iii) halogenated acids (e.g. herbicides) and oxidation products of halogenated compounds.

(i) In the atmosphere tetra-alkylated lead (R_4Pb) compounds are readily converted to R_3Pb^+ and R_2Pb^{2+} on attack by OH. In solution, R_4Pb seems to oxidise through a range of paths (some photochemical) to R_3Pb^+.

(ii) In addition to the production of hydroxymethylhydroperoxides in rainwater as outlined in Section 5.5, they are readily produced in the gas phase from the ozonolysis of naturally occurring alkenes (i.e. the pinenes and terpenes described in Section 2.1). One can write the reactions for the formation of hydroxymethyl-hydroperoxide as

$$CH_2CR_2 + O_3 \rightarrow CH_2OO + R_2CO \tag{R7.5}$$

$$CH_2OO + H_2O \rightarrow HOCH_2O_2H \tag{R7.6}$$

and the equation for formation of the *bis*-hydroxymethylhyd-roperoxide as

$$HOCH_2O_2H + CH_2O \rightarrow HOCH_2O_2CH_2OH \qquad (R7.7)$$

(iii) It may be true today that the halogenated acids have the greatest potential to cause damage. Chlorofluorocarbons are gradually being replaced by less stable compounds that react in the troposphere and pose less of a threat to the ozone layer. These compounds often react in the atmosphere to produce halogenated acids. HCFC-124 will typically degrade following the patterns for hydrocarbons that was seen in Section 6.5:

$$CF_3CFClH + OH \rightarrow CF_3CFCl + H_2O \qquad (R7.8)$$

$$CF_3CFCl + O_2 \rightarrow CF_3CFClO_2 \qquad (R7.9)$$

$$CF_3CFClO_2 + NO \rightarrow CF_3CFClO + NO_2 \qquad (R7.10)$$

$$CF_3CFClO \rightarrow CF_3CFO + Cl \qquad (R7.11)$$

The highly fluorinated acetylfluoride (CF_3CFO) dissolves in water droplets and reacts to give trifluoroacetic acid:

$$CF_3CFO + H_2O \rightarrow CF_3COOH + HF \qquad (R7.12)$$

The ultimate fate of the acid remains uncertain.

Effects on surface waters
The increase in acidity of rainfall within Europe is parallelled by an increase in the acidity of surface waters. These changes have been responsible for the most dramatic effect of increasing acidity, namely the destruction of aquatic life. Long-term changes in the acidity of some lakes within Europe are shown in Fig. 7.13. These changes over long periods of time have been established through paleolimnological studies of the faunal record in the lake sediments.

The changes in lakewater pH took place over rather restricted geographical areas, which do not necessarily correspond to the areas of highest acidic flux. A reason for this is that many lakes and the soils that surround them act as buffers with sufficient capacity to prevent large changes in pH. However, large areas of Scandinavia possess poorly buffered soils derived from lime deficient rocks that are resistant to the weathering processes. Such soils are poor sinks for the deposited hydrogen ions. Similar geology is found in some parts Scotland and in

North America, where surface water acidification is also a problem.

The decreasing pH of ground waters in these regions has caused concern about the increase in the concentrations of aluminium, copper and cadmium in drinking water. High concentrations of aluminium have been loosely associated with *senile dementia*; although copper and cadmium are toxins they have not been conclusively associated, as yet, with prevalent modern diseases.

The most serious concern from the declining pH of surface waters is its

Fig. 7.13. Long-term changes in the acidity of some lakes in Europe derived from paleolimnological studies (from Renberg and Battarbee (1990)).

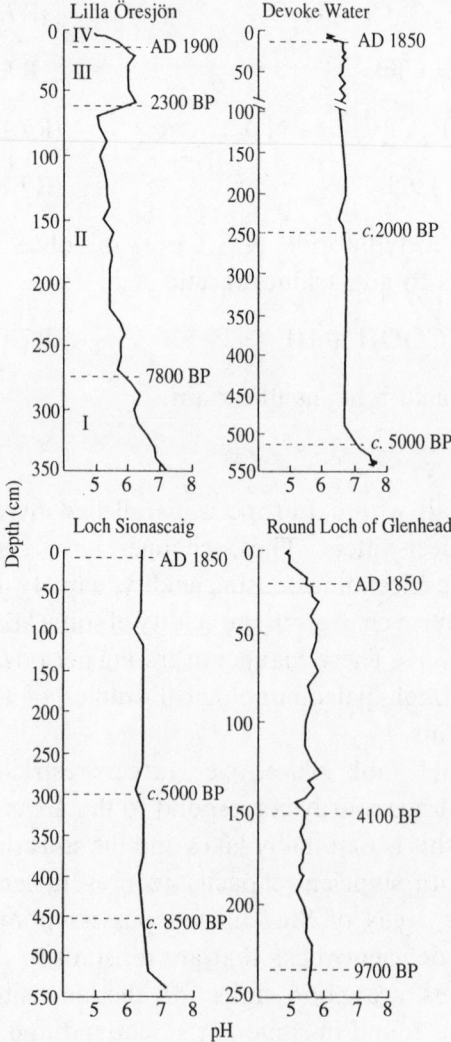

connection with the decrease in fish and amphibian populations of lakes in both Europe and in North America. A pH range 6–9 is desirable to support a good fish stock and below pH 5 lakes do not seem to have healthy fish. The rainbow trout seems to be an especially sensitive species. Records suggest that salmon and trout catches have been declining since late last century, in the areas of Norway currently worst affected by acidification. It is as though the recent acidification has dramatically increased changes that may have already been underway. Earlier changes could have been agricultural or transport of low-level industrial emissions of air pollutants last century. Most of the current acidification is due to atmospheric pollution and especially that from power generation.

The aluminium ion is thought to be an important cause of the toxicity of acid waters for fish. The toxicity of aluminium is dependent on pH, reaching a maximum for the brown trout at pH 5.0. Both the aluminium and the hydrogen ion affect the gill, by altering the active uptake of sodium ions. Aluminium hydroxide precipitates at the gill filament and clogs it with mucus. Acid waters enhance this toxic effect of aluminium because the lower pH values increase the aluminium solubility. The toxicity of aluminium and H^+ is lessened if calcium is relatively abundant in the water because it can reduce the permeability of the gill membrane to Na^+ and H^+. The high aluminium concentration in acid waters has also been implicated in *senile dementia*, by virtue of the interaction of aluminium with neuron fibrillary tangles in the brain.

A further impact of low pH is the damage that it does to fish eggs. At pH 4.1 newly fertilised salmon eggs die within two days. A result is that many lakes in Scandinavia and North America are now without fish or at least have very depleted populations dominated by older fish.

A significant fraction of the precipitation in some regions affected by acid rain arrives as snow. The melt water from this snow is often extremely acidic. If the melting is rapid then the water comes off quickly and flows over the surfaces of rocks. This means that it has little time for contact with soils, which would have helped to increase pH. Similar pulses of highly acidic water may be observed after heavy storms. These events have been responsible for many fish being killed. Often fish can tolerate low pH, but are killed by rapid changes.

The acid that is present in snow melt is concentrated by yet another process. As a snow pack melts, the acidic components are removed preferentially. Some 80% of the solute load is removed by the first 20% of the melt water. This can easily mean a factor of four increase in acid concentration. If this takes place on the first warm days of spring it can lead to a massive pulse of acid that does enormous damage to aquatic life. The exact method through which preferential removal of solutes occurs is not well understood. It could be that the solutes are located on the grain boundaries within the snow, as the salt solutions are expelled from the ice or that there is some kind of chromatographic concentration effect as the melt water percolates down through the snow column.

Two types of solutions to the acid rain problem have been proposed. One involves treatment at the site of damage. This would include liming lakes and their catchments with alkaline material. However, this method is not always very successful. The problem may also be treated at source, by removing the pollutants from the emissions. As coal-burning power stations are the largest source of acid pollutants, there is a strong movement to control their emissions. Here the solutions usually amount to desulphurisation of flue gases. This can remove 80–90% of the sulphur from the gas stream, but increased capital and running costs are of the order of 10–20%. Different systems of firing can lead to considerable reduction in the levels of nitrogen oxide emission, but this is usually only achieved at the cost of a loss in fuel efficiency. However, the desulphurisation of flue gases may have the additional advantage of reducing NO_x emissions.

7.8   Global pollution

The acid rain problem is a regional one, as the important effects are found about a thousand kilometres from the source. Residence times for the components of acid rain are of the order of several days. This means rapid rates of removal and makes the effects regional rather than global. However, atmospheric pollutants with longer residence times have a global effect.

The best known of these is carbon dioxide. Atmospheric carbon dioxide concentrations have been measured for more than a hundred years, but large errors are associated with the earliest measurements. Even so there was sufficient interest for some scientists of the late nineteenth and early twentieth centuries, to suggest that CO_2 variations exerted an important control over the Earth's climate. Historical and contemporary analyses, plus gas samples from ice cores show a very substantial increase that has occurred over the last few centuries (see Fig. 7.14). Most of this increase, which currently runs at about 1.8% per annum, has been attributed to the combustion of fossil fuels. However, the quantities burnt are so large that one expects an increase nearly double that which is observed. Much of this additional 'lost carbon dioxide' appears to have been absorbed by the oceans. It has also been suggested that the carbon lost from the soil during deforestation is a further large source of CO_2 in the atmosphere, but the magnitude of this source is more difficult to estimate than the combustion source. The best modern budgets suggest a further missing sink for about $1.2\,\text{Pg(C)}\,\text{a}^{-1}$, which could be explained in terms of increased terrestrial carbon sequestration from fertilisation effects.

The new estimates available suggest that combustion has probably released carbon dioxide at a rate of about $6\,\text{Pg(C)}\,\text{a}^{-1}$ in the 1990s, while the input from changing land-use patterns and deforestation might best be placed at about $1.6\text{--}2.0\,\text{Pg(C)}\,\text{a}^{-1}$. This should be compared with the amount of carbon in various reservoirs shown in Fig. 7.15. There is an additional $0.1\,\text{Pg(C)}\,\text{a}^{-1}$ from industries such as cement manufacture. The quantities released through burning fuel are but a small fraction of those present in the various natural reservoirs, but they do represent a significant fraction of the natural flux to and from the atmosphere.

Increases in fossil fuel consumption have been slightly more than 4% per annum in the period 1950–1980. It has become important to predict the rate of increase in carbon dioxide emission for future decades. This, naturally enough, depends largely on the way in which fossil fuels are used. High use scenarios place emissions at $20\,\text{Pg(C)}\,\text{a}^{-1}$ by the middle of the next century, but most predictions are about half this.

The principal effect of increasing the carbon dioxide concentration will be an increase in the temperature of the atmosphere. As shown in Chapter 1, the Earth is maintained at a temperature above that expected from radiative equilibrium through the greenhouse effect. Carbon dioxide and other trace gases are transparent to incoming solar radiation, but absorb outgoing long-wave radiation. In this way they are able to raise the

Fig. 7.14. Long-term changes in atmosphere carbon dioxide concentrations.

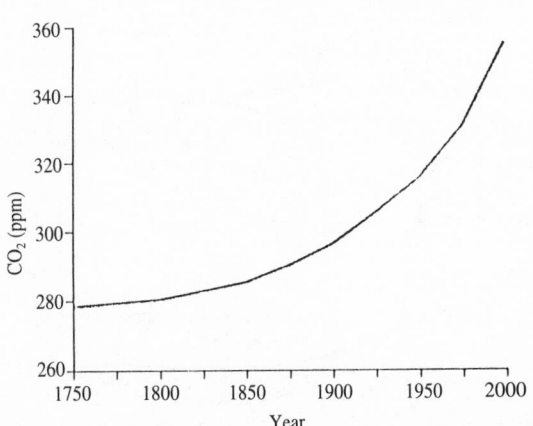

surface temperature. The issue is more complex than it first appears, because, at some wavelengths, virtually all outgoing radiation is absorbed and any further addition of carbon dioxide would have no effect. However, at other wavelengths the absorbtion is much weaker and increases become important. Alternatively we could imagine the gas layer responsible for emission being pushed higher in the atmosphere as carbon dioxide concentration increased. At higher altitudes it would be colder and following Boltzmann's Law, emit less effectively; thus the world would become warmer.

It has not been easy to predict the increases in temperature that take place with given increases in atmospheric carbon dioxide concentration. The difficulty arises because of the great complexity of the global circulation and its influence on climate. It is also difficult to compute changes in cloudiness and the role of the oceans in controlling temperature.

Some models suggest that a doubling in the level of CO_2 would give rise to an increase of 1.5–4.5 °C, although heating at the poles could be much greater. One might argue that global warming would be beneficial. Milder winters would save fuel and agriculture would benefit from longer growing seasons. It is also possible that plants can grow more effectively at high carbon dioxide concentrations, although this supposition requires that other factors do not limit growth. However, climatic change might also give rise to drier conditions and a rise in insect and microbial pests.

Fig. 7.15. The global carbon cycle (from *EPRI Journal*, October/November 1993). The reservoir sizes in parentheses are in Gt(C) and fluxes are in Gt(C) a^{-1}.

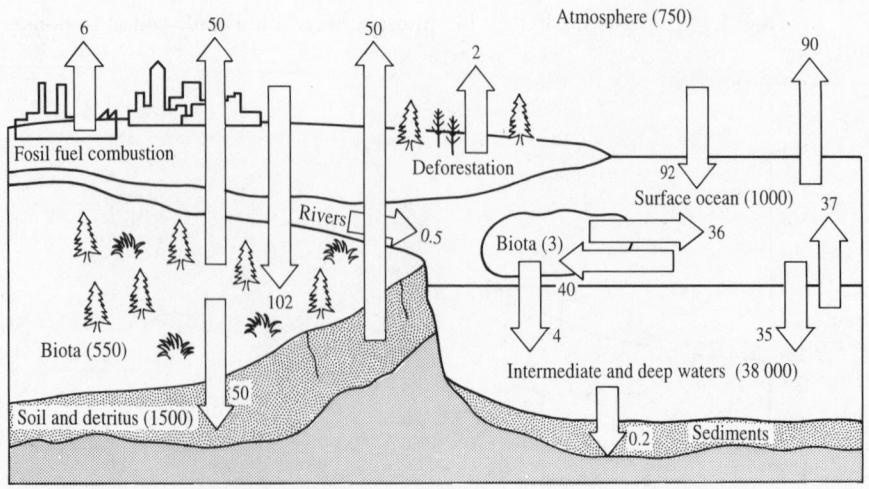

Furthermore, warmer polar waters may induce a melting of the polar ice caps or shelves and raise the sea level. Direct effects on human health are likely to be small.

Any discussion of the greenhouse gases now has to broaden beyond carbon dioxide. A range of other radiatively active gases are playing an increasing role in the absorption of outgoing radiation (see Table 7.9). Despite the accumulation of relatively small quantities of stable gases such as methane and the halocarbons, especially the chlorofluoroalkanes, these are very efficient absorbers and play a role in the greenhouse effect. Halocarbons are effective because they are efficient absorbers in the infra red region. They now show measurable increases in concentration as illustrated in Fig. 7.16. These data comes both from recent measurements and, over the longer time scale, from air samples trapped in glacial ice. Interestingly methane, which has increased rapidly in concentration since about 1800, is showing slower rates of increase in the 1990s. This may be due to a decreased leakage from pipelines (especially in Siberia), less coal mining or lowered methane release from the Arctic permafrost.

There are also aspects of human activity that have probably served to lower global temperature. Sulphur emissions to the atmosphere, especially the northern hemisphere increased markedly from the mid-twentieth century onwards. This sulphur is converted to sulphate aerosol, which is able to cool the Earth in two ways. First, it acts as condensation nuclei for clouds that reflect radiation, while in cloud-free areas it can back-scatter incoming radiation to space. This latter effect may be the stronger of the two. Evidence for the role of sulphur dioxide in reducing the global temperature rise may be found in observations that suggest that the less

Table 7.9. *Properties of greenhouse gases*

Gas	Concentration (ppb) (1990)	Trend (a^{-1})	Lifetime (a)	Radiative forcing contribution 1765–1980	1980–90
CO_2	353000	0.4%	100–250	61%	55%
CH_4	1720	1.0%[1]	8	17%	11%
CFC-11	0.28	4%	65	12%	24%
CFC-12	0.48	4%	120		
N_2O	310	0.3%	150	4%	6%
$C_2Cl_3F_3$	0.06	10%	90		
CH_3CCl_3	0.16	4%	6		

[1]Trend rapidly decreasing in 1990s.

polluted southern hemisphere has warmed more than the northern hemisphere after the 1950s. Growth rates in the emissions of anthropogenic sulphur emissions were much smaller in the southern hemisphere for the first half of the present century.

Although most attention has been devoted to the role of various gases in changing climate, it is also possible that increasing concentrations of trace gases have altered the chemistry of the atmosphere. The best known of these changes may be the increased amounts of ozone observed over continental atmospheres during the last century. Ozone concentrations could have doubled in the last hundred years. This is usually taken to be the result of enhanced photochemistry because of higher concentrations of reactive anthropogenic hydrocarbons in the atmosphere.

The hydroxyl radical exerts a major control over the trace gas

Fig. 7.16. The temporal increase in atmosphere concentration of a number of radiatively active gases (Wuebbles, 1993).

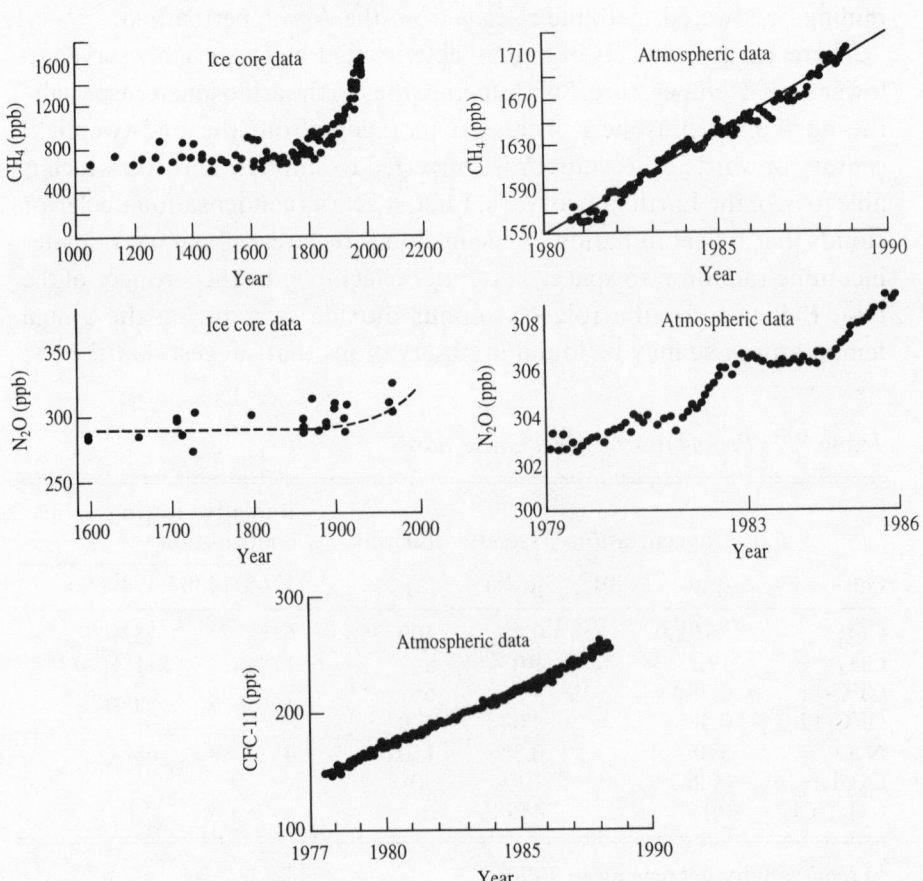

chemistry of the atmosphere. Carbon monoxide represents a major sink for the hydroxyl radical, so variations in carbon monoxide and hydroxyl concentration can be closely linked. We have no long-term records of OH concentration, but background carbon monoxide concentrations were rising through the 1980s. However, current data suggest decreasing concentrations. This could be caused by increases in OH concentration (the result of enhanced UV radiation) or, which is more likely, a decline in the anthropogenic emissions of CO.

7.9   Further reading

Brimblecombe, P. (1990) Composition of museum atmospheres, *Atmospheric Environment* **B24**, pp 1–8.

Hewitt, C. N. and Sturges, W. T. (1993) *Global Atmospheric Chemical Change*, Elsevier, London and New York.

Howells, G. (1989) *Acid Rain and Acid Waters*, Ellis Horwood, Chichester.

Hulme, M. and Jones, P. D. (1994) Global climate change in the instrumental period, *Environmental Pollution* **83**, 23–36.

Leslie, G. B.and Lanau, F. W. (1992) *Indoor Air Pollution Problems and Priorities*, Cambridge University Press, Cambridge.

Mason, B. J. (1992) *Acid Rain*, University Press, Oxford.

Treshow, M. and Anderson, F. K. (1991) *Plant Stress from Air Pollution*, John Wiley & Sons, Chichester.

Wellburn, A. (1994) *Air Pollution and Climate Change*, Longman, Harlow, Essex.

WHO (1987) *Air Quality Guidelines for Europe*, World Health Organization, Copenhagen.

8

○ ○

The upper atmosphere

Early chemists expected the heavier gaseous elements of the air to sink to the bottom of the atmosphere, but, as we saw in Chapter 1, the composition remains constant with height. However, this is true only of the lower atmosphere, where turbulent eddies keep the constituents well mixed. Above about 100 km the atmospheric density becomes so low that molecular diffusion begins to dominate over eddy diffusivity and the gases separate gravitationally. Therefore the lighter gases are found higher up and hydrogen is the dominant component in the outermost reaches of the atmosphere. Above the turbopause, which marks the upper limit of turbulent mixing, different gases have different scale heights because the molecular weight term in equation (1.5) is no longer the mean molecular weight of the entire atmosphere. Calculations need to be done separately for each gas. The heavier the gas the smaller the value of the scale height. The separation of the constituent gases explains the name 'heterosphere' that is often applied to the part of the atmosphere that lies above the turbopause. This distinguishes it from the well mixed 'homosphere' that lies below. Profiles for some gases at high altitude are shown in Fig. 8.1.

Separation of gases as a function of their molecular weight means that, at high altitude, the relative proportions of various gases can change quite dramatically. This should be remembered when looking at the gases in the succeeding sections of this chapter. Gas pressure or amount of gas can fall with height, yet its concentration in mole fraction or ppm terms could be increasing dramatically.

8.1 Ozone and the chemistry of the stratosphere

The chemistry of ozone in the upper atmosphere has received a great deal of attention during the past decade, because of the vital role that it plays in shielding organisms on the Earth from damaging UV radiation. This interest has been heightened by observations that current emissions from anthropogenic activities have changed the concentrations of ozone in the upper atmosphere. While ozone is an extremely important component of the stratosphere, it is found in surprisingly small amounts. If all the ozone in the Earth's atmosphere, most of which is found in the stratosphere, were brought to ground level it would constitute a layer of pure ozone only 3 mm thick. Our improved understanding of the chemistry of ozone in the upper atmosphere makes any simple treatment difficult because

Fig. 8.1. The composition of the upper atmosphere. The exobase lies in the region delimited by the dotted lines. Its exact height varies with solar activity.

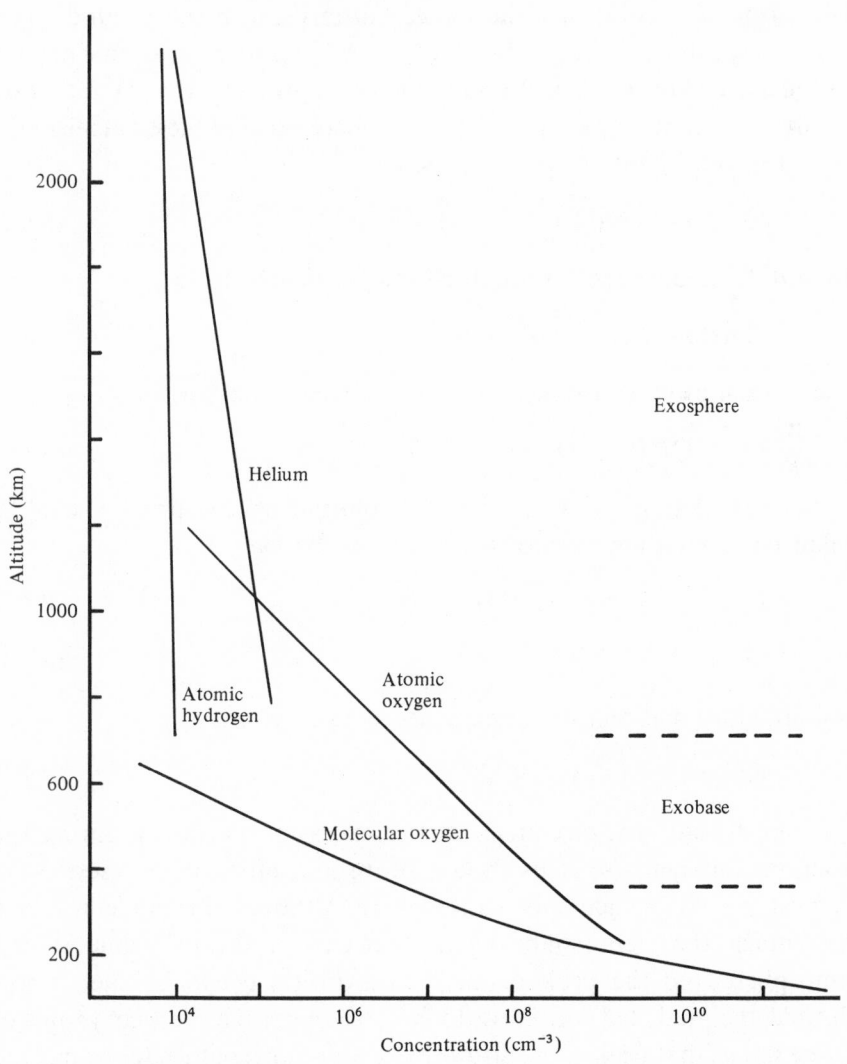

many hundreds of reactions are now invoked in the chemistry of ozone. Here we will try to examine some reactions that give most insight into the chemistry of the upper atmosphere. Its sensitivity to human activity at the surface of the Earth will be a special focus.

The upper atmosphere is dry and cold so the production of OH from water no longer plays the major role that it does in the troposphere. The chemistry at these altitudes more usually involves atomic oxygen. Ozone formation is a photochemical process, involving UV radiation of less than 242 nm and oxygen:

$$O_2 + hv \rightarrow O(^3P) + O(^3P) \qquad \text{(R8.1)}$$

The terms ^3P signify that the oxygen atoms are in the ground state. Photodissociation at wavelengths less than 175 nm yields atomic oxygen in an excited state, $O(^1D)$, although such short-wavelength UV radiation is only important higher up in the atmosphere because these wavelengths cannot penetrate into the stratosphere:

$$O_2 + hv \rightarrow O(^3P) + O(^1D) \qquad \text{(R8.2)}$$

The $O(^1D)$ is quenched through collisional deactivation:

$$O(^1D) + M \rightarrow O(^3P) + M \qquad \text{(R8.3)}$$

The oxygen atom can now react with molecular oxygen:

$$O_2 + O(^3P) + M \rightarrow O_3 + M \qquad \text{(R8.4)}$$

The production of ozone by this photochemical process can be balanced against the reactions that destroy ozone:

$$O_3 + hv \rightarrow O_2 + O(^1D) \qquad \text{(R8.5)}$$

$$O_3 + O \rightarrow 2O_2 \qquad \text{(R8.6)}$$

and an additional removal process for oxygen atoms:

$$O + O + M \rightarrow O_2 + M \qquad \text{(R8.7)}$$

The 'third body' M, of course, carries excess energy away during the reaction. The reaction steps shown above are sufficient to describe the chemistry of an 'oxygen only' stratosphere. Although this model will give very much the right shape to the vertical profile of ozone in the atmosphere and the peak ozone concentration occurs at the correct altitude, the predicted concentrations are too high. The vertical profile of ozone shows that peak concentrations occur at lower altitudes nearer the poles (see Fig 8.2). In the 1960s it became evident that more reactions were required if the chemistry of ozone were to be modelled correctly. A set of analogous reactions involving hydrogen-, nitrogen- and chlorine-containing species is possible:

$$OH + O_3 \rightarrow O_2 + HO_2 \qquad\qquad\qquad (R8.8)$$

$$HO_2 + O \rightarrow OH + O_2 \qquad\qquad\qquad (R8.9)$$

which sums to

$$O_3 + O \rightarrow 2O_2 \qquad\qquad\qquad (R8.10)$$

Similar reactions can be written for other species:

$$NO + O_3 \rightarrow O_2 + NO_2 \qquad\qquad\qquad (R8.11)$$

$$NO_2 + O \rightarrow NO + O_2 \qquad\qquad\qquad (R8.12)$$

and

$$O_3 + Cl \rightarrow O_2 + ClO \qquad\qquad\qquad (R8.13)$$

$$ClO + O \rightarrow O_2 + Cl \qquad\qquad\qquad (R8.14)$$

Each of these reaction pairs sums in a way that involves the destruction of

Fig. 8.2. Partial pressure of ozone as a function of height. When plotted in these units, the 'ozone layer' becomes very clear. The height of the ozone layer varies with latitude and season.

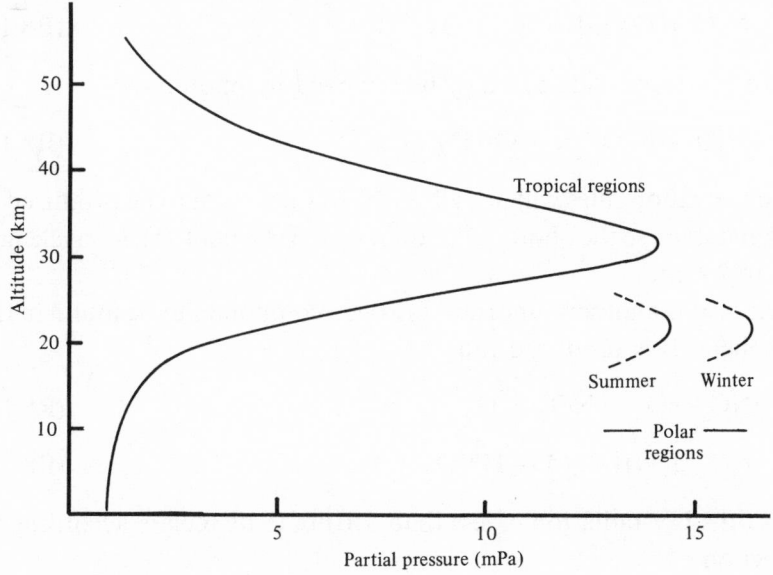

ozone and atomic oxygen while restoring the OH, NO or Cl molecules.
Thus, these processes are catalytic and each of the reactive H, N and Cl
species may be responsible for the destruction of many ozone molecules.
When these extra reactions that destroy ozone are considered, much
better models of the ozone layer are possible.

We can now see why so much concern has been expressed about trace
contaminants in the stratosphere. A single pollutant molecule can be
responsible for the destruction of many ozone molecules, by initiating a
chain of reactions. Therefore, it is important to account for the chemistry
of a range of minor constituents of the upper atmosphere.

Nitrogen chemistry of the stratosphere
Aircraft operating at high altitude are a direct source of nitrogen oxides in
the stratosphere, but the number of aircraft at these altitudes is relatively
small. The main source of NO in the stratosphere is nitrous oxide
produced through bacterial action at the surface of the Earth. There is
evidence of other sources and, in recent times, indications that vehicles
with catalytic converters are effective emitters. Nitrous oxide is stable and
well mixed within the troposphere, so it gradually crosses into the
stratosphere, where it can react with the $O(^1D)$ produced by sources such
as the photolysis of ozone or, at high altitudes, bimolecular oxygen:

$$N_2O + O(^1D) \rightarrow 2NO \tag{R8.15}$$

or

$$N_2O + O(^1D) \rightarrow N_2 + O_2 \tag{R8.16}$$

However, nitrous oxide can also be removed by photolysis:

$$N_2O + hv \rightarrow N_2 + O(^1D) \tag{R8.17}$$

Spin conservation rules will not allow the oxygen atom to be produced in
the ground state, so the photolysis requires energetic photons of wavelength
less than 260 nm.

Nitric acid is another important nitrogen compound to be found in the
stratosphere. It is produced thus:

$$NO + O_3 \rightarrow NO_2 + O_2 \tag{R8.18}$$

$$NO_2 + OH + M \rightarrow HNO_3 + M \tag{R8.19}$$

and absorbs UV radiation of less than 330 nm to dissociate according to
the reaction

$$HNO_3 + hv \rightarrow OH + NO_2 \qquad \text{(R8.20)}$$

The ozone layer removes sufficient of the shorter wavelength UV radiation for nitric acid to have a photochemical lifetime of about ten days at the tropopause. This is long enough for much of the HNO_3 to be transferred across the boundary and removed in rainfall. This is the major means by which oxidised nitrogen is lost from the stratosphere.

Chlorine chemistry of the stratosphere
The main natural source of chlorine in the stratosphere is probably methyl chloride, but this accounts for only 25% of that currently transported across the tropopause. Concern about the impact of anthropogenic chlorine compounds on the chemistry of the stratosphere began in the early 1970s. Then, it was realised that the chlorofluorocarbons (CFCs) used as aerosol propellants and refrigerants had become widely distributed through the troposphere. There was no obvious mechanism for destruction of these highly stable compounds in the lower atmosphere. Thus, it was not difficult to guess that, rather like the stable N_2O, the chlorofluoromethanes would be transported to the stratosphere. The most important chlorofluoromethanes, $CFCl_3$ (Freon-11) and CF_2Cl_2 (Freon-12), absorb UV radiation in the 190–220 nm range, which results in the photodissociation reactions

$$CFCl_3 + hv \rightarrow CFCl_2 + Cl \qquad \text{(R8.21)}$$

$$CF_2Cl_2 + hv \rightarrow CF_2Cl + Cl \qquad \text{(R8.22)}$$

The free chlorine atom can now react with ozone in the catalytic manner noted above in (R8.13). Chlorine monoxide produced on reaction with ozone does not always react with atomic oxygen (as in reaction (R8.14)). Occasionally it can form a dimer:

$$ClO + ClO \rightarrow Cl_2O_2 \qquad \text{(R8.23)}$$

which will photolyse to yield chlorine atoms

$$Cl_2O_2 + hv \rightarrow 2Cl + O_2 \qquad \text{(R8.24)}$$

which will re-enter the ozone-destruction cycle shown in equations (R8.13) and (R8.14).

Chlorine monoxide can also interact with nitrogen compounds:

$$ClO + NO \rightarrow Cl + NO_2 \qquad \text{(R8.25)}$$

$$ClO + NO_2 + M \rightarrow ClNO_3 + M \qquad \text{(R8.26)}$$

The chlorine nitrate ($ClNO_3$) may be decomposed by UV radiation or by reaction with atomic oxygen:

$$ClNO_3 + O \rightarrow O_2 + ClO + NO \tag{R8.27}$$

The reaction of ClO and NO_2 (R8.26) is important because they effectively remove the nitrogen and chlorine species from the cycles that destroy ozone. They essentially sequester the important ozone destroying nitrogen and chlorine species. If these reactions were not to be considered in modelling studies, then the impact of chlorine compounds on the ozone layer might be overestimated. Chlorine in the stratosphere also reacts with methane or hydrogen to give hydrogen chloride:

$$Cl + CH_4 \rightarrow HCl + CH_3 \tag{R8.28}$$

$$Cl + H_2 \rightarrow HCl + H \tag{R8.29}$$

Some of this hydrogen chloride will react in the stratosphere to return the free chlorine atom:

$$HCl + OH \rightarrow H_2O + Cl \tag{R8.30}$$

but ultimately it has to cross the tropopause to be removed from the atmosphere by rainfall as hydrochloric acid.

Other sources of chlorine in the stratosphere include the chlorine from ammonium perchlorate–aluminium solid rocket propellants. However, chlorine compounds are not always anthropogenic. A whole suite of halogenated compounds of natural origin find their way into the stratosphere, but these are currently dwarfed by the human contribution.

Bromine chemistry of the stratosphere
Organobromine compounds are increasingly used in many industrial applications. Halon compounds (e.g. $CBrF_3$ and $CBrClF_2$) found in fire extinguishers are typical of the anthropogenic bromine compounds that now diffuse their way to the stratosphere. These enter into ozone-destruction cycles:

$$CBrF_3 + hv \rightarrow CF_3 + Br \tag{R8.31}$$

$$Br + O_3 \rightarrow BrO + O_2 \tag{R8.32}$$

$$Cl + O_3 \rightarrow ClO + O_2 \tag{R8.33}$$

$$BrO + ClO \rightarrow Cl + Br + O_2 \tag{R8.34}$$

which sums to

$$2O_3 \rightarrow 3O_2 \qquad\qquad\qquad (R8.35)$$

General changes

Ozone's photochemical origin means that variations in the solar irradiance will alter its production rate. There are also variations due to circulation changes in the atmosphere. In the tropics, seasonal variation is small in ozone, whereas the variation at high latitude can be large as is shown in Fig. 8.2. Under current ozone-depletion regimes, in the stratosphere, the seasonal changes can be very large with major reductions in the southern polar spring.

The ozone layer is very susceptible to changes. Changes may have occurred after nuclear explosions that have introduced large quantities of NO_x into the stratosphere or when nearby supernovae have irradiated the upper atmosphere with cosmic rays. During the fall of the Tunguska meteor over Siberia in 1908, it is calculated that some 30 Tg of NO were deposited between 10 and 100 km altitude. This is about five times the amount of oxidised nitrogen species ($NO + NO_2 + HNO_3$) contained within the entire stratosphere. The presence of such a large quantity of NO_x could have greatly affected the ozone layer. Records of atmospheric transparency (in the UV) made between 1909 and 1911 suggest that the atmosphere was recovering from unusually low levels of ozone.

8.2   Stratospheric aerosols and the ozone hole

Early models of the change in ozone brought about by human activities wavered between high and low estimates of damage, but they converge on suggestions that, on average, depletion would be a few percent per decade or so. Current models still suggest modest, yet still worrying, depletions in middle latitudes. However, while these models were still being refined, massive ozone depletion was discovered over the Antarctic in spring (Fig. 8.3). This was far too large to be explained in terms of what was known of stratospheric chemistry.

This discovery prompted the realisation that explanations require us to consider the chemistry of cloud droplets in a region of the atmosphere with very little water. The stratosphere is also very cold, so water can only form in the presence of soluble materials that depress the freezing point. Some sampling programmes to examine the composition of aerosol droplets of the lower stratosphere began in the late 1950s. There proved to be high concentrations of sulphate anions and early studies suggested that the aerosols were largely ammonium sulphate. Such aerosols formed a layer that was named after the air chemist C. E. Junge. However, it has become increasingly evident that much of the ammonia in early samples arose from contamination. Current analyses suggest that the sulphate aerosol of the Junge layer is sulphuric acid. Early speculations as to the source of the sulphate aerosol in the stratosphere centred on Junge's idea that the aerosol arose from the chemical transformation of H_2S and SO_2 that had crossed the tropopause in the tropics. It was also realised that volcanic eruptions could eject gases into the stratosphere and that was therefore another possible source for the aerosol.

The thin clouds that result from these aerosols have long been observed, but the volume of the droplets was always taken to be so small that heterogeneous chemistry could be neglected. Interest in this aerosol layer previously centred on the way in which it affected the radiation balance of the stratosphere. Changes in this would ultimately feed back into shifts in the Earth's temperature. Such speculation was not without support, because observations that explosive volcanic eruptions in the past had led to pronounced changes in climate were well documented. Benjamin Franklin had commented on the prevalence of a universal fog and the weakness of solar radiation in 1783, when the volcano Laki erupted in Iceland. The years that followed the eruption were particularly cold ones. The process was repeated with the eruption of Hekla in 1821. Here the gloom was claimed to be responsible not only for a change in

Fig. 8.3. Ozone in October over Halley Bay, Antarctica (from Stolarski (1993)).

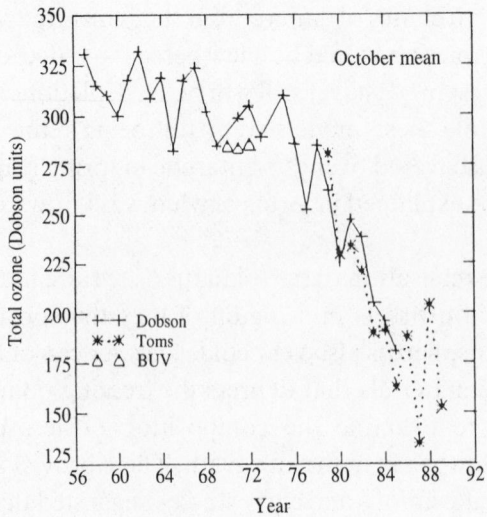

climate, but also for inducing the production of the novel *Frankenstein*, that year, by a bored and housebound Mary Shelley.

In more recent times the eruptions of Agung, Mount St Helens, El Chichon and Pinatubo have been studied to ascertain the effect that they have on the stratospheric aerosol concentration and on climate. Agung had a clear influence on stratospheric temperatures in the years that followed its eruption in 1963, but the global effects were hardly catastrophic. Mount St Helens, while a spectacular eruption, had little effect and even the eruption of El Chichon in a more equatorial location led to changes that were smaller than predicted.

It is possible that even quite small volcanic eruptions that proceed largely unnoticed can affect the aerosol load in the stratosphere. A continuous contribution to the aerosol layer might arise from this source. The importance of volcanoes is also seen in measurements made on the composition of polar ice. Dated Antarctic ice cores show acid bands that correlate with the dates of major eruptions. In the centre of the Antarctic continent, the contribution of maritime sources to the sulphate in snow is small. In such a remote and clean location that is also favoured by transport of materials down from the stratosphere, it may be possible to detect changes in stratospheric sulphate levels by examining ice cores. High concentrations of sulphate are found in snow that corresponds to the dates of the eruption of Krakatoa (1883) and Agung (1963).

Volcanic contributions to the stratosphere take the form of dust and gaseous components. Dust is thought to settle out relatively quickly. The production of particles through gas– particle conversion processes is likely to maintain higher levels of stratospheric aerosol for much longer periods. The injection of large quantities of SO_2 will generate an aerosol through the reaction

$$OH + SO_2 \rightarrow HOSO_2 \qquad\qquad (R8.36)$$

which could significantly deplete the concentrations of OH in the upper atmosphere. The subsequent reactions of HSO_3 are not clearly known, but possible steps are

$$HOSO_2 + O_2 \rightarrow HO_2 + SO_3 \qquad\qquad (R8.37)$$

$$H_2O + SO_3 \rightarrow H_2SO_4 \qquad\qquad (R8.38)$$

which leads to formation of the ubiquitous sulphuric acid aerosol. Oxidation of volcanic SO_2 in the stratosphere is thought to be slow and residence times of as much as a year have been proposed. Sulphuric acid formed in this way can then condense with the small amount of water vapour, at high altitude, to give aerosol droplets. However, the fact that small insoluble particles are found in such droplets adds support to the notion that condensation occurs onto Aitken-sized particles (see Fig. 4.4 for the definition of Aitken-sized particles).

Volcanoes are just one possible source of sulphur gases in the stratosphere. Biogenic carbon disulphide has been suggested as an important source of sulphur. Measurements indicate that the CS_2 concentrations in the stratosphere are extremely low, but it is likely to be transferred across the tropopause rather than other inorganic sulphur compounds such as H_2S and SO_2. These seem to be too reactive to escape from the troposphere. In Fig. 8.4 are shown profiles for some sulphur compounds in the atmosphere. The particulate sulphur of the Junge layer can be clearly seen in a graph that expresses concentrations as a mole fraction. This is convenient because it allows us to plot both the gaseous and the condensed phases on the same diagram. It is apparent from the diagram that carbonyl sulphide should also be important in the chemistry of stratospheric sulphur compounds. It can be oxidised to sulphuric acid comparatively easily through photochemical processes or via oxygen-atom attack:

Fig. 8.4. The profiles of various sulphur compounds with height. Note the sulphuric acid layer in the stratosphere. (Data from Georgii and Meixner (1980), Whitten *et al.* (1980) and Turco *et al.* (1981)).

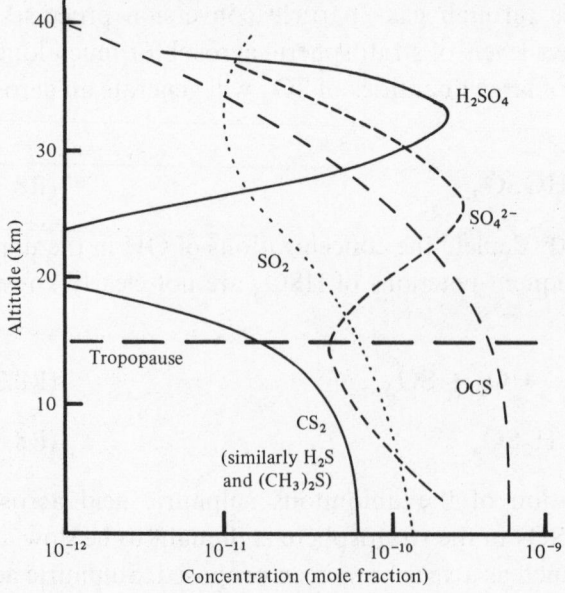

$$COS + hv \rightarrow CO + S \tag{R8.39}$$

$$S + O_2 \rightarrow SO + O \tag{R8.40}$$

$$COS + O \rightarrow CO + SO \tag{R8.41}$$

$$SO + O_2 \rightarrow SO_2 + O \tag{R8.42}$$

$$SO + O_3 \rightarrow SO_2 + O_2 \tag{R8.43}$$

Some estimates place the anthropogenic contributions to the COS flux into the atmosphere as high as 50%. The burning and processing of fossil fuels is probably the largest source. If this is correct then such contributions may greatly affect the concentrations of stratospheric aerosol and therefore climate.

Other sources of anthropogenic particulate material in the stratosphere can be more direct. For instance, it has been calculated that the aluminised rubber binder fuel used in the space shuttle will give $AlCl_3$, $Al_2O_3 \cdot HCl$ and Al_2O_3 aerosols and make a significant contribution to the number of particles with radii less than 0.01 μm.

Polar stratospheric cloud chemistry
Earliest interest in the stratospheric aerosol was limited to its production and physical properties until polar stratospheric clouds (PSCs) began to be seen as relevant to ozone depletion. The discovery of the Antarctic ozone hole prompted scientists to consider the way in which ozone might be removed so effectively in the polar regions.

It had long been recognised that the formation of chlorine nitrate (R8.26) effectively acts as a reservoir for potentially ozone-depleting nitrogen and chlorine entities. Polar stratospheric clouds seemed to offer potential surfaces for rapid heterogeneous reactions that could convert reservoir species into active chlorine. Two reactions were soon found to be efficient:

$$ClONO_{2(g)} + HCl_{(s)} \rightarrow Cl_{2(g)} + HNO_{3(s)} \tag{R8.44}$$

$$ClONO_{2(g)} + H_2O_{(s)} \rightarrow HOCl_{(g)} + HNO_{3(s)} \tag{R8.45}$$

These processes push chlorine compounds into the gas phase, but retain the nitrogen in the solid phase. The solid nitric acid is to be found in the stratosphere as nitric acid trihydrate or NAT ($HNO_3 \cdot 3H_2O_{(s)}$), which is the basis of what are now called Type 1 polar stratospheric clouds.

The chlorine molecules and hypochlorous acid can be photolysed to chlorine atoms:

$$Cl_{2(g)} + hv \rightarrow 2Cl_{(g)} \tag{R8.46}$$

$$HOCl_{(g)} + hv \rightarrow Cl_{(g)} + OH \tag{R8.47}$$

The chlorine now enters the destructive cycle that turns out to be very fast at low temperatures. One can also see that reaction (R8.49) involves a square dependence on ClO making it sensitive to chlorine concentration:

$$2[O_3 + Cl \rightarrow O_2 + ClO] \tag{R8.48}$$

$$ClO + ClO + M \rightarrow Cl_2O_2 + M \tag{R8.49}$$

$$\frac{Cl_2O_2 + hv \rightarrow 2Cl + O_2}{2O_3 \rightarrow 3O_2} \tag{R8.50}$$
$$\tag{R8.51}$$

Clouds play a role in ozone depletion, which explains why the injection of volcanic material into the stratosphere can be so important. The volcanogenic aerosol will act as an additional site for ozone depleting processes.

Heterogeneous processes are particularly effective in the polar spring. Low temperatures of the winter ensure the formation of polar stratospheric clouds. Nitric acid incorporated into cloud particles can slowly sediment leading to a denitrification of the stratosphere. In early spring, light reaches the poles, but it will be too weak to photolyse gas-phase nitric acid. These processes result in a low gas phase abundance of the nitrogen oxides. The reactions that convert chlorine into inactive reservoir species will be ineffective, so the fast catalytic cycle (R8.48)–(R8.51) will become very important, leading to formation of the spring ozone hole over Antarctica.

8.3   The upper atmosphere

The success of early radio transmissions across the Atlantic convinced physicists that the atmosphere of the Earth had a conducting layer at an altitude of some 80 km. This layer became known as the ionosphere and was probed with radio waves from below. Later, studies were extended with radio-sounders carried aloft by rockets, or gas samplers that collected material for chemical analysis. Early work established the well-developed layered structure shown in Fig. 8.5. The layers originally detected by radio reflection also proved to be useful in describing chemical structure. However, remember that there is, in reality, a single ionospheric layer and the boundaries represent changes in the gradient of electron density that refract the impinging radio waves.

The first studies showed pronounced diurnal and long-term changes in the structure of the ionosphere. Long-term changes usually follow the solar sunspot cycle. The changes not only affect the reflection of radio waves, but also alter the concentrations of various species in the upper atmosphere. The electrons in the ionosphere are produced by photo-ionisation.

$$N_2 + h\nu(< 79.6\,\text{nm}) \rightarrow N_2^+ + e^- \tag{R8.52}$$

$$O_2 + h\nu(< 102.6\,\text{nm}) \rightarrow O_2^+ + e^- \tag{R8.53}$$

$$O + h\nu(< 91.1\,\text{nm}) \rightarrow O^+ + e^- \tag{R8.54}$$

Reaction (R8.53) dominates the ionising process in the E-region, while (R8.54) and to a lesser extent (R8.52) are important in the lower part of the F-region (Fig. 8.5).

Reactions (R8.52) and (R8.53) can proceed by dissociative ionization at shorter wavelengths:

$$N_2 + h\nu(< 66.2\,\text{nm}) \rightarrow N^+ + N + e^- \tag{R8.55}$$

$$O_2 + h\nu(< 51.0\,\text{nm}) \rightarrow O^+ + O + e^- \tag{R8.56}$$

Fig. 8.5. Electron densities and some positive ion concentrations in the ionosphere. The letters indicate the D, E, F1 and F2 regions of the ionosphere.

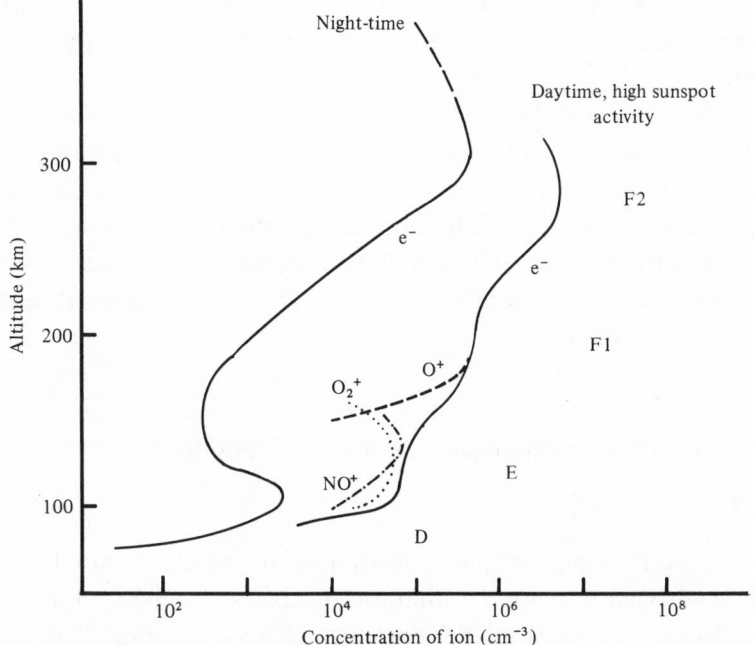

Above 100 km the photo-ionisation is largely caused by extreme UV radiation that is absorbed by O, O_2 and N_2. It just happens that the wavelength of the intense hydrogen Lyman-α radiation (121.57 nm) occurs at a wavelength for which the absorption cross section of O_2 is low, and that it is too long to be absorbed by O and N_2. This allows Lyman-α radiation to penetrate into the mesosphere where it contributes to the formation of the D-region. There are also small contributions from cosmic ray bombardment at slightly lower altitudes, where the atmosphere becomes less tenuous. However, magnetic shielding means that cosmic radiation is only important at polar latitudes. The production of electrons at night presents a problem because cosmic rays are not an adequate source for this. Night-time ionisation has often been attributed to a downward flux of protons and UV night-glow (radiation from excited species in the upper atmosphere).

Photodissociation of molecular oxygen coupled with slow recombination in the thermosphere accounts for the dominance of atomic oxygen above about 200 km (Fig. 8.1).

$$O_2 + h\nu(< 242.2\,\text{nm}) \rightarrow O + O \tag{R8.57}$$

or yields an excited oxygen atom:

$$O_2 + h\nu(< 174.9\,\text{nm}) \rightarrow O(^1D) + O(^3P) \tag{R8.58}$$

Nitrogen is photodissociated by being excited into some energetic pre-dissociation states by radiation of shorter wavelength in the range 80–100 nm.

The photo-ionisation and impact ionisation account for the initial production of charged and atomic species in the upper atmosphere. However, the composition of the ionosphere is sometimes rather different from that expected simply from primary production processes. One striking observation is that, although NO is a minor neutral species, NO^+ is a major ion in the E-region. Production, in this region, occurs through the ion–atom interchange:

$$O^+ + N_2 \rightarrow NO^+ + N \tag{R8.59}$$

A further production mechanism also acts as a sink for N_2^+:

$$O + N_2^+ \rightarrow NO^+ + N \tag{R8.60}$$

Lower in the D-region, photo-ionisation of the readily ionised nitric oxide by hydrogen Lyman-α radiation produces the NO^+ ion. Ion chemistry in this region is further complicated by hydration. The high dipole moment of water molecules allows them to pick up positive or negative charges with comparative ease. At 80 km altitude, much of the positive charge is associated with water-derived ions or hydrates. These are species such as H_3O^+, $H_5O_2^+$ and $H_7O_3^+$, which are observed and likely to be formed in reactions such as

$$O_2^+ + O_2 + M \rightarrow O_4^+ + M \tag{R8.61}$$

$$O_4^+ + H_2O \rightarrow O_2 + O_2^+ \cdot (H_2O) \tag{R8.62}$$

$$O_2^+ \cdot (H_2O) + H_2O \rightarrow H_3O^+ + OH + O_2 \tag{R8.63}$$

However, the most distinguishing feature of the D-region is probably the formation of negative ions, e.g.

$$O_2 + e^- + O_2 \rightarrow O_2^- + O_2 \tag{R8.64}$$

The production of electrons and ions is balanced by the loss processes in a quasi-steady state ionosphere. Reaction (R8.64), which is a 'three-body attachment', is typical of these processes. The loss processes usually involve reduction of the photo-electron to thermal energies followed by ion–electron recombination or electron attachment. Other characteristic reactions are

$$NO^+ + e^- \rightarrow N + O \tag{R8.65}$$

which is a dissociative recombination and

$$O^+ + e^- \rightarrow O + h\nu \tag{R8.66}$$

which is a radiative recombination.

8.4 Sporadic processes in the ionosphere

The low number density of atomic and molecular entities in the upper atmosphere means that the chemistry is relatively easily perturbed. Radio and rocket studies of the E-region have shown that there is always a residual degree of ionisation at night, in spite of the rapid recombination of major molecular ions. Night-time layers consist of metal ions that survive because of a slow rate of recombination. Sodium is the most abundant metal (with a column density of 5×10^9 neutral atoms cm^{-2}) and a powerful night-glow line that allowed its detection as early as 1929. Mean peak concentrations of about 5×10^3 atoms cm^{-3} occur near 90 km altitude. The presence of a persistent night-time ionisation and the appearance of thin sporadic layers has suggested that the metals in the upper atmosphere are ablated from meteors.

Some principal reactions of meteoritic sodium in the upper atmosphere are now understood. Sodium can be photo-ionised or gain a charge by atom–ion interchange, to form Na^+ that dominates above 100 km. The major chemical transformation of neutral sodium occurs via oxidation:

$$Na + O_3 \rightarrow NaO + O_2 \tag{R8.67}$$

$$NaO + O_3 \rightarrow NaO_2 + O_2 \tag{R8.68}$$

The superoxide (NaO_2) is particularly important at night because concentrations at altitudes below 90 km can build up in the absence of the photolysis process

$$NaO_2 + h\nu \rightarrow O_2 + Na \tag{R8.69}$$

During the day the bicarbonate is likely to be a major component:

$$NaO + H_2O \rightarrow NaOH + OH \tag{R8.70}$$

$$NaOH + CO_2 + M \rightarrow NaHCO_3 + M \tag{R8.71}$$

The chemistry of other important metals in the upper atmosphere, such as K, Li, Ca, Mg and Fe, is less well understood. There is also speculation that the accumulation of meteoritic metals in the winter polar stratosphere might interact with the chlorine-induced destruction of ozone. In particular, alkaline salts of these metals could react with nitric acid and remove it from catalytic cycles.

If a small meteor can affect the chemistry of the ionosphere, it is not surprising that the effects of anthropogenic activities are also detectable. When Skylab was launched, it was the first large booster to operate in the upper portion of the ionosphere (190 km). Electron densities were lowered over a radius of 1000 km around the flight path of the rocket. This electron-hole can be understood by considering the fact that some 1.2×10^{31} molecules of water and hydrogen were released during the ionospheric portion of the flight. Oxygen ions (O^+) are normally removed from the ionosphere through reaction with nitrogen and oxygen:

$$O^+ + O_2 \rightarrow O_2^+ + O \tag{R8.72}$$

$$O^+ + N_2 \rightarrow N_2^+ + O \tag{R8.73}$$

but the reactions involving hydrogen or water are some thousand times faster. This leads to considerable reduction in electron concentration through the reactions

$$O^+ + H_2O \rightarrow H_2O^+ + O \tag{R8.74}$$

$$O^+ + H_2 \rightarrow OH^+ + H \tag{R8.75}$$

followed by a range of dissociative processes:

$$e^- + H_2O^+ \rightarrow H_2 + O \tag{R8.76}$$

$$e^- + H_2O^+ \rightarrow OH + H \tag{R8.77}$$

$$e^- + OH^+ \rightarrow O + H \tag{R8.78}$$

8.5   Policy and change

Depletion of the ozone layer has now been recognised as one of the most dramatic changes that human activities have caused to the atmosphere. We have seen in earlier sections how catalytic cycles serve to amplify the effects of quite small amounts of material in the stratosphere. The CFCs have been identified as the most important contributors to the change. The Antarctic ozone hole shows the largest changes, but even at lower latitudes there have been changes of a few percent per decade that are larger than was expected from early models.

It is not entirely clear how depletion of ozone in the stratosphere will be parallelled by increases in UV radiation at ground level. A decrease in ozone will lead to greater UV transparency in the stratosphere, but other changes can counterbalance some of this. Over the polluted continents, increases in fine particles in the troposphere have raised the opacity of the troposphere towards UV radiation. Nevertheless, these uncertainties cannot conceal the need for swift action to prevent further deterioration of the ozone layer.

It was fortunate that, in the 1970s, before confirmation of destruction of the ozone layer by CFCs, many nations reduced their use of these compounds (Table 8.1). The USA banned non-essential use of CFCs in 1978. In 1980 the European Community issued a Directive that placed a ceiling on production and achieved a 35% reduction in CFCs by the early 1980s. However, these early gains were offset by a widening global market. The Vienna Convention (1985) began international discussions that eventually led to the Toronto Protocol (1987), which required specific international action on five CFCs and three other halons. It did this by freezing and subsequently reducing production. It also recognised the difficulties that developing nations would face by delaying restrictions, in those countries, for a decade. The protocol was intended to lead to a 50% reduction in global CFC production. There was a danger that this would not be met, so some countries (initially the USA and UK) committed themselves to 85% reductions. Germany and Sweden then opted for virtual elimination. However, Japan, with a huge dependence on CFC-113 as a cleaning agent in its electronics industry, was reluctant to take such steps. India, Brazil and the People's Republic of China were heavily committed to the mass production of refrigerators. They argued that the industrialised nations had benefited from cheap, convenient CFCs, so these nations should bear the costs that developing countries would have to meet in using environmentally less destructive compounds.

During the London Conference, Australia, the European Community, Canada and New Zealand pushed for earlier dates for the elimination, but nations with huge commitments to CFCs (e.g. India, the USSR and the USA) thought 2000 the earliest date possible. There was also agreement to take care over the use of CFC replacement and put aside money to aid developing nations in adopting ozone-friendly substances.

The replacements for CFCs would ideally be non-toxic, non-flammable, stable and inexpensive. The substitutes found so far do not match all of these criteria. The general approach has been to design molecules that are not completely halogenated and thus have hydrogen atoms bound to the carbon atoms, which makes them more reactive towards tropospheric OH. These compounds have often been called HCFCs. They break down readily to carbon dioxide and halogenated acids, but these also cause

Table 8.1. *Ozone-depleting substances and potential replacements. ODP, the ozone depleting potential, is given relative to CFC-11 and CFC-12.*

Compound	Abbreviation	ODP	Lifetime (a)	Use
$CFCl_3$	CFC-11	1.0	65	Blowing agent, refrigerant
CF_2Cl_2	CFC-12	1.0	120	Refrigerant
$C_2F_3Cl_3$	CFC-113	0.8	90	Cleaning agent
$C_2F_4Cl_2$	CFC-114	1.0	180	Blowing agent, refrigerant
C_2F_5Cl	CFC-115	0.6	380	Refrigerant
CF_2BrCl	Halon-1211	3	21	
CF_3Br	Halon-1301	10	110	Fire extinguishant
$C_2F_4Br_2$	Halon-2402	?	26	Fire extinguishant
CF_3CHCl_2	HCFC-123	0.02	1.7	To replace CFC-11, Halon 1211
$CHClF_2$	HCFC-22	0.055	15.8	To replace CFC-11
CF_3CH_2F	HCFC-134a	0.00	15.6	To replace CFC-12
$CF_3CF_2CHCl_2$	HCFC-225ca	0.025	2.8	To replace CFC-113
CH_3CCl_3			6.5	
CCl_4			50	

some environmental problems, as discussed in Section 7.7, (R7.7)–(R7.11). Another approach would be to design fluorocarbons, as fluorine does not participate in the ozone-destruction cycles.

The HCFCs are generally more complex to make because of their lower stability. In addition they have to be larger, and thus more expensive, molecules to have sufficiently high boiling points to be useful in refrigeration and air-conditioning systems. Some replacement compounds are likely to be used only briefly, until more satisfactory compounds have been evaluated. Despite this effort, damage to the ozone layer will persist for many years because of the long lifetime of the halogenated species that have already been released.

8.6　Further reading

Chamberlain, J. W. and Hunten, D. M. (1987) *Theory of Planetary Atmospheres*, Academic Press, Orlando.

Hamill, P. and Toon, O. B. (1991) Polar stratospheric clouds and the ozone hole, *Physics Today* **44**(12) 34–42.

Kaye, J. A. and Jackman, C. H. (1993) Stratospheric ozone change, in Hewitt, C. N. and Sturges, W. T. *Global Atmospheric Chemical Change*, Elsevier, London.

Peters, T (1994) The stratospheric ozone layer, *Environmental Pollution*, **83**, 69–79.

Plane, J. M. C. (1991) The chemistry of meteoritic metals in the Earth's upper atmosphere, *International Reviews in Physical Chemistry*, **10**, 55–106.

Rees, M. H. (1989) *Physics and Chemistry of the Upper Atmosphere*, Cambridge University Press, Cambridge.

9

○ ○

The atmospheres of the planets and their evolution

Thus, far this book has concerned itself with the composition and chemistry of the Earth's atmosphere. Little has been said of the origin and evolution of atmospheric composition, although it might have appeared natural to discuss this at the beginning of the book. However, it is probably easier to view the origin and evolution of the Earth's atmosphere against the background of the composition of planetary atmospheres. This is because studying the atmospheres of other planets not only provides us with some remarkably interesting atmospheric chemistry, but also offers important clues about the origin of the atmosphere of the Earth.

9.1   The stability of planetary atmospheres

Many planets, moons and other bodies are known, but any list shows that few have atmospheres. Table 9.1 gives the properties of some named bodies, starting with the planets closest to the sun. The moons are listed next to their parent planets. In keeping with many similar tabulations, the word 'none' may appear in the column that describes the atmospheres of the bodies. Obviously, 'none' may simply mean that the atmosphere is too tenuous to detect. The planet Mercury and the Moon have very thin atmospheres. If most of the other bodies in the table had atmospheres as thin as these then they would not have been detected yet.

Strictly speaking, it might be best to call the most tenuous atmospheres exospheres. Here the mean free path of the molecules in such thin atmospheres exceeds the scale height. This means that there is no equilibrium between the translational degrees of freedom possessed by the gas molecules.

It may be that bodies that now have only exospheres never had substantial atmospheres or that they lost them. The question of retention or stability of an atmosphere is an important one. After all, a planetary atmosphere is a container for gases that lacks a wall, so the retention of gases is primarily the result of gravitational attraction. This leads to a potentially leaky container.

The earliest view of the escape of gases from planetary atmospheres was thermal evaporation. This process was examined in detail in the pioneering work of Jeans early this century. It is based on the idea that molecules in the exosphere undergo ballistic trajectories, so any with velocities higher than the escape velocity of the planet can be lost into space. The escape velocity from a planet, v_e, is given by the equation

$$v_e = (2MG/r_e)^{0.5} \tag{9.1}$$

where M is the mass of the planet, G the gravitational constant $(6.67 \times 10^{11}\,\mathrm{N\,m^2\,kg^{-2}})$ and r_e the radius of the planet at the base of the exosphere (the exobase, see Fig. 8.1). The most probable velocity of molecules of compound i, v_i, is given by the equation

$$v_i = (2RT/M_i)^{0.5} \tag{9.2}$$

where R is the gas constant, T is the absolute temperature and M_i is the molecular weight of the compound.

As a rule of thumb, planets will tend to lose their atmosphere, over the time equivalent to the age of the solar system, if the molecular velocity v_i is greater than about a quarter of the escape velocity. Escape occurs because the distribution of molecular velocities is such that many exceed the mean. A further notional relationship of interest is to argue that exospheric temperature declines with the square root of distance from the sun. Thus the further away a planet is, the lower the exospheric temperature and the greater the likelihood of it retaining its atmospheric constituents. These ideas are summarised in Fig. 9.1, which overlays two graphs. The first plots the bodies in the solar system as a function of their distance from the sun and escape velocity, while the second gives the velocity of various molecules as a function of temperature. Bodies with substantial atmospheres are shown as dots while those without atmospheres, or with at the most exospheres, are plotted as open circles. We can see from this diagram that the notion of thermal escape is a very useful one for predicting whether a planet will have an atmosphere or not. Cold massive bodies (i.e. those with high gravitational attraction and therefore high escape velocities) are the ones that retain an atmosphere.

However, it has been realised in recent years that non-thermal loss

Table 9.1. *Data for some named bodies in the solar system.*

The list of satellites is far from complete, only the larger ones being included. Similarly, just a few of the asteroids are tabulated. The data are taken largely from Kaufmann(1979), Henderson-Sellers (1983) and the Chemical Rubber Co *Handbook of Physics and Chemistry 1974–5*. The semi-major axis is the distance from the sun for planets and asteroids, and the distance from the parent planet for moons.

Name	Mass (kg)	Gravity (m s^{-2})	Radius (km)	Semi-major axis (km)	Surface temperature (K)	Albedo	Atmosphere	Symbol
Mercury	3.3(23)	3.95	2440	5.79(7)	440	0.12	Tenuous	☿
Venus	4.9(24)	8.87	6050	1.08(8)	730	0.59	CO_2	♀
Earth	6.0(24)	9.78	6380	1.50(8)	288	0.39	N_2, O_2	⊕
Moon	7.6(22)	1.62	1740	3.84(5)	274	0.11	Tenuous	☾
Mars	6.4(23)	3.73	3400	2.28(8)	218	0.15	CO_2	♂
Phobos	9.6(15)		14×10	9.38(3)		0.05	None	
Deimos	2.0(15)		8×6	2.35(4)		0.05	None	
Ceres			510	4.14(8)		0.06	None	
Vesta	$\simeq 0.2(20)$		274	3.53(8)		0.24	None	
Jupiter	1.9(27)	23.2	71 900	7.78(8)		0.44	H_2, He	♃
Amalthea			135×78	1.82(5)		0.05	None	
Io	8.9(22)	1.79	1820	4.22(5)	110	0.63	Tenuous	J_I
Europa	4.8(22)		1565	6.71(5)		0.64	None	J_{II}
Ganymede	1.5(23)		2640	1.07(6)		0.43	None	J_{III}
Callisto	1.1(23)		2420	1.88(6)		0.17	None	J_{IV}
Saturn	5.7(26)	8.77	60 330	1.43(9)		0.46	H_2, He	♄
Dione	1.1(21)		560	3.77(5)		0.5	None	
Rhea	2.5(21)		765	5.27(5)		0.6	None	
Titan	1.4(23)		2580	1.22(6)		0.2	N_2	S_{VI}
Iapetus	1.9(21)		730	3.56(6)			None	

Uranus	8.6(25)	9.46	26 150	2.87(9)	0.56	H$_2$, He	♅
Titania	1.2(21)		800	4.38(5)	0.23	None	
Oberon	1.2(21)	13.7	810	5.86(5)	0.18	None	
Neptune	1.0(26)		24 750	4.50(9)	0.51	H$_2$, He	N$_I$
Triton	5.7(22)		1 700	3.55(5)	0.4	Trace	P
Pluto	9(21)		1 550	5.90(9)	?	Trace	
Charon			750	1.7(4)	?	None	

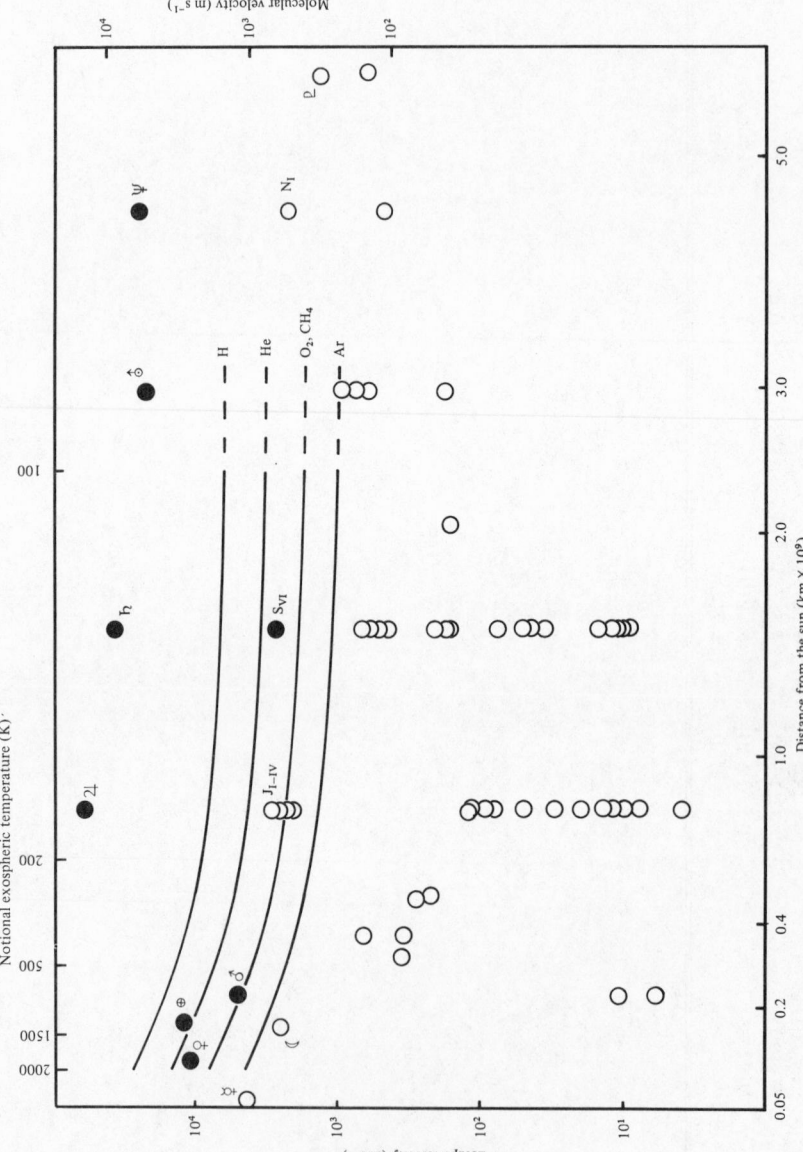

Fig. 9.1. Escape velocities for various bodies in the solar system and their distance from the sun. The bodies without substantial atmospheres are represented by empty circles. The curves indicate molecular velocities for some likely gases as a function of temperature. The temperature scale at the top represents notional exospheric temperatures. Most of the bodies that have dense atmospheres have conditions that limit molecular escape. See Table 9.1 for explanation of symbols.

mechanisms represent a control on the atmospheres of planets. These processes, excepting photodissociation involve ions. Ions usually have difficulty escaping from magnetic fields associated with planets unless the field lines are open or they are so far from the planet that they can be picked up by the solar wind. Alternatively, chemical energy can be converted into sufficient kinetic energy to allow a neutral atomic product to escape. Dissociative recombination:

$$O_2^+ + e^- \rightarrow O^* + O^* \tag{R9.1}$$

$$OH + e^- \rightarrow O + H^* \tag{R9.2}$$

or ion–neutral reactions of the type

$$O^+ + H_2 \rightarrow OH^+ + H^* \tag{R9.3}$$

are typical. The asterisked atoms carry away much of the kinetic energy. Energy can also come from an incident electron:

$$e^{-*} + N_2^+ \rightarrow N^* + N^* \tag{R9.4}$$

photon:

$$h\nu + O_2 \rightarrow O^* + O^* \tag{R9.5}$$

or fast particle:

$$Na + S^{+*} \rightarrow Na^* + S^+ \tag{R9.6}$$

9.2 Surface boundary exospheres

Mercury

Non-thermal mechanisms become necessary because the escape mechanism described by Jeans cannot completely explain the absence of atmospheres on other bodies. Heavy molecules should be retained even at high temperatures such as are found on the planet Mercury. The theory of thermal escape of gases would lead us to expect an exosphere of argon (600 K), let us say, derived from radiogenic decay of potassium to be retained. This is because such an atmosphere would be estimated to survive more than 10^{20} a! This is far longer than the age of the solar system of some 4.5 Ga.

However, Table 9.2 suggests that Mercury does not have an extensive argon atmosphere despite the liklihood of there being a strong source of argon. Argon atmospheres are good absorbers in the UV, but are poor emitters. This means that they would soon reach very high temperatures

(> 4000 K) that would favour thermal loss of argon. Alternatively, argon is susceptible to ionisation and being swept away by the solar wind.

The atmosphere of Mercury was studied by instruments carried on board the spacecraft Mariner 10. Encounters with the planet provided much of our early knowledge of its atmospheric composition by detecting helium and hydrogen. Both these gases are lost very effectively by thermal and non-thermal mechanisms, so we need to postulate sources to maintain the low steady state concentrations observed. Helium can arise from α-particles produced radiogenically from the uranium and thorium decay series. Both hydrogen and helium can be captured from the solar wind. The dominance of helium in the atmosphere could well be a product of less effective removal of helium compared with hydrogen. Several other minor sources have been proposed: volatiles carried in by meteoritic material, out-gassing of volatiles from the planetary interior and through production by reactions between the solar wind and the planetary surface. Subsequent telescopic observations have revealed an extensive sodium and potassium exosphere.

The Moon

Manned landings on the Moon have provided an excellent opportunity to place sensitive equipment on the Moon to detect the presence of extremely low concentrations of gas. The measurements were difficult because of contamination by out-gassing from the landing vehicle during the lunar day. However, helium and argon are important constituents of the lunar exosphere. A smaller quantity of molecular hydrogen is predicted in theoretical calculations. Neither oxygen nor carbon dioxide were detected, so their concentrations are at least an order of magnitude lower than those of the inert gases. Some argon is ^{36}Ar, implying a non-radiogenic source for argon, but most arises from the decay of ^{40}K. Spatial and temporal variations of argon in the lunar atmosphere are probably related to seismic activity. The most likely control on helium in the lunar

Table 9.2. *Composition of surface boundary exospheres (atoms cm^{-3})*

Element	Mercury	Moon
Hydrogen	8	<10
Helium	4 500	$<2 000$
Argon	<500	1 600
Sodium	26 000	40
Potassium	500	10

exosphere is the solar wind that emanates from the sun.

When we look at the bodies with tenuous atmospheres, we do find situations that are always in accord with the simple notion of thermally controlled escape. However surface boundary exospheres are atmospheres in which weak sources are overwhelmed by effective loss mechanisms. Once such loss processes dominate, they readily reduce atmospheres to their minimal exospheric state.

9.3   Tenuous colder atmospheres

There are some atmospheres, which, although tenuous, have a high enough number density for molecular collisions to be frequent. These atmospheres are often associated with smaller bodies in the solar system that have difficulty retaining well-developed atmospheres, but the atmosphere is often maintained by evaporation from surface ices or frosts of methane, carbon monoxide, sulphur dioxide or water.

Io

Io, an active and fascinating satellite, is an inner Galilean moon of Jupiter. This moon is the only body, apart from the Earth, where active vulcanism is currently observed. The satellite has long been thought to have an atmosphere because asymmetrical sodium and potassium clouds were detected near the moon (Fig 9.2). It is possible that these alkali metal atoms are sputtered from salts on the surface of the planet by particles such as protons with a few thousand electron-volts of energy or effectively by heavy nuclei:

$$Na + S^{+*} \rightarrow Na^* + S^+ \tag{R9.7}$$

Besides the alkali metal cloud, there is an incomplete toroidal cloud of

Fig. 9.2. Hydrogen and sodium clouds associated with Io.

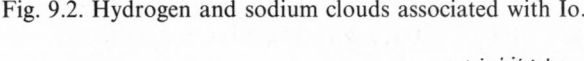

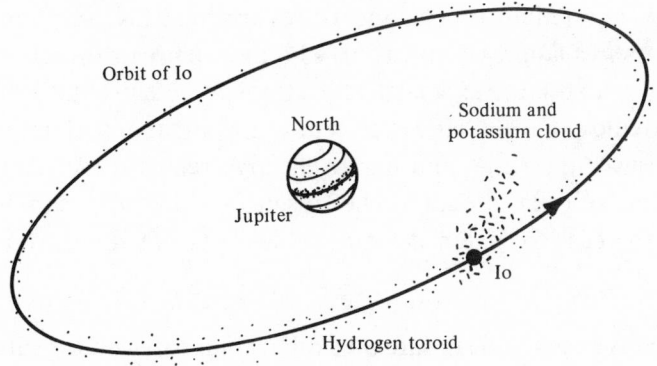

hydrogen atoms surrounding Jupiter (Fig 9.2). This is associated with the orbit of Io and could be maintained by a hydrogen source of $10^{10}\,cm^{-2}\,s^{-1}$ at the surface.

The active vulcanism on Io is understandable considering the enormous tidal stresses imparted by its proximity to Jupiter. Volcanic plumes extend as far as 300 km from the moon. Extreme UV spectra showed the presence of ionised atomic oxygen and sulphur. Lighter gases would be lost from a body as small as Io due to the presence of intense particle radiation and magnetic fields from Jupiter. However, a heavier gas such as sulphur dioxide can be retained. The geological features observed on the satellite are strongly indicative of sulphur melting or the volatilisation of compounds such as sulphur dioxide.

We can imagine the sulphur dioxide being photodissociated under UV irradiation in a variety of ways:

$$SO_2 + hv \rightarrow SO + O \tag{R9.8}$$

$$SO_2 + hv \rightarrow S + O_2 \tag{R9.9}$$

the sulphur monoxide could photodissociate or react:

$$SO + SO \rightarrow S + SO_2 \tag{R9.10}$$

If we regarded the main loss of oxygen as reaction with sulphur compounds:

$$O_2 + SO \rightarrow SO_2 + O \tag{R9.11}$$

$$O_2 + S \rightarrow SO + O \tag{R9.12}$$

and a surface that is an effective sink for sulphur then it might be removed sufficiently rapidly for oxygen to be an important constituent of the atmosphere.

Another interesting question is that of whether the atmosphere is in equilibrium with surface deposits of sulphur dioxide. The Jovian moon has a noon temperature of 130 K, with even higher values in certain hot spots (290 K), and night-time temperatures around 95 K. Sulphur dioxide vapour pressures could be up to 0.014 Pa at a temperature of 130 K, so pressures near to this are expected if the atmosphere is in equilibrium with sulphur dioxide as a solid. However, it now appears that surface pressures are much lower than this, hinting at effective removal mechanisms for volcanogenic sulphur dioxide. The absence of a polar cap has also weakened the case for extensive surface deposits of sulphur dioxide.

Ganymede
Ganymede, like the Galilean satellites other than Io, contains substantial quantities of water as ice. One could postulate an atmosphere in

equilibrium with ice and expect water vapour to be present. It has also been noted that water vapour in the atmosphere could lead to the formation of oxygen atmospheres through photolysis and loss of hydrogen. Oxygen, being heavier and less easily condensable than water, could favour the development of more substantial atmospheres on 'icy' bodies yielding oxygen pressures around a few tenths of a pascal. However the *Voyager* encounter set the oxygen pressure on Ganymede many orders of magnitude lower than this.

Titan

The presence of an atmosphere on Titan, the largest moon of Saturn, was predicted by Jeans early this century on the basis that it would be large and cold enough to retain methane in substantial quantities. Methane was detected in the spectrum of Titan in the 1940s and for many years it was the only satellite known to possess an atmosphere. The Voyager 2 encounter of Saturn provided an excellent opportunity for studying the atmosphere of Titan. Virtually featureless clouds cover the moon. Most models propose a layer of photochemical haze above the cloud deck (Fig. 9.3). The composition of Titan's atmosphere is given in Table 9.3.

Methane makes up between 1% and 6% of the composition of the atmosphere of the satellite. Models for the development of Titan's atmosphere assume that the body initially had no atmosphere; the gases being trapped as chlathrate compounds within ice (e.g. methane as $CH_4 \cdot 7H_2O$). During subsequent radiogenic heating, methane was released to the surface. Ammonia would also have been incorporated initially as ice-chlathrate compounds, but while methane can exist as a gas at current temperatures, ammonia would be present as a condensed phase. This makes it difficult to explain the presence of large amounts of nitrogen in the atmosphere of Titan. The most obvious source of the gas would be photolysis of gas phase ammonia into nitrogen and hydrogen followed by escape of hydrogen into a toroidal cloud about Saturn. One way of explaining the atmospheric nitrogen is to imagine that Titan was warm enough in some early epoch of its history for gas-phase ammonia to exist in equilibrium above an ammonia ocean. Over some 100 Ma this would have been photolytically oxidised to nitrogen. Another possibility is that the moon accreted at very low temperatures ($< 60 \, \mathrm{K}$), which would have allowed the incorporation of argon and molecular nitrogen into icy chlathrates. In such a picture, the volatilisation of nitrogen and argon from the ice would be expected to show the cosmic ratio of argon : nitrogen of $1:11$, which is not too far from the values currently suggested in some determinations. Certainly, the large quantities of argon proposed for the present atmosphere are too large to be explained by radioactive decay of ^{40}K.

The presence of methane in the atmosphere and a distinct planetary surface has led to much speculation about the possibility of extensive organic chemistry in the atmosphere of Titan. It seems a more likely candidate for complex organic chemicals than the atmospheres of the Jovian planets. A further interesting piece of speculation has been the idea that Titan may resemble the Earth in that methane may act in the same way as water does on the Earth and be found in all three phases. Many calculations suggest that there must be large quantities of methane on the surface of Titan. The methane may be present as an ocean, although partial pressure considerations would require this ocean to contain substantial amounts of nitrogen, ethane and propane.

Chemistry on Titan is dominated by photolysis of CH_4 in a nitrogen atmosphere:

$$CH_4 + h\nu \rightarrow CH_2 + H_2 \tag{R9.13}$$

$$CH_2 + CH_3 \rightarrow C_2H_4 + H \tag{R9.14}$$

Fig. 9.3. The temperature–pressure relationship and cloud composition for Titan.

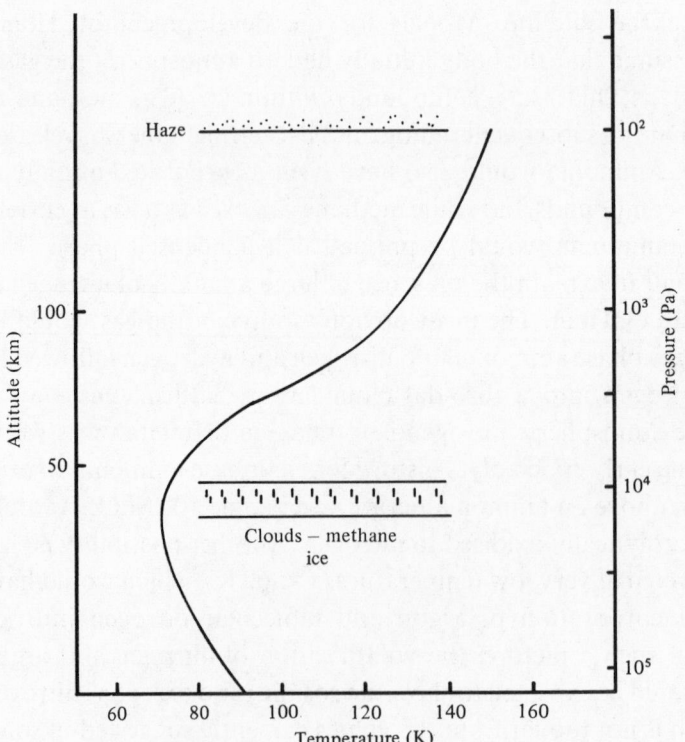

This ethene provides a source of both ethyne (acetylene) and ethane:

$$C_2H_4 + hv \rightarrow C_2H_2 + H_2 \tag{R9.15}$$

$$C_2H_4 + H + M \rightarrow C_2H_5 + M \tag{R9.16}$$

$$C_2H_5 + H \rightarrow CH_3 + CH_3 \tag{R9.17}$$

$$CH_3 + CH_3 + M \rightarrow C_2H_6 + M \tag{R9.18}$$

The circuitous method of production involving the third body M is probably required to carry off extra energy. Ethane appears to dominate over ethyne in the atmosphere of Titan, see Table 9.3. Ethene is reactive and, more importantly, is readily photolysed, so its concentrations remain low.

We could easily imagine the ethyl radical (C_2H_5) from reaction (R9.16) and the methyl from (R9.17) reacting to give a ready source of propane:

$$C_2H_5 + CH_3 + M \rightarrow C_3H_8 + M \tag{R9.19}$$

Acetylene acts as a catalyst for the reactive cycles because it absorbs

Table 9.3. *Atmospheric composition (Pa) of some colder planets and moons*

	Pluto	Ganymede	Io	Titan	Triton
Nitrogen	1(sh)			133 000	1.6
Argon				13 000	
Hydrogen				400	
Carbon monoxide	0.1–1(sh)				0.002
Methane	0.07			1 300	Trace
Ethane				3	
Ethene				0.1	
Ethyne				0.5	
Propane				0.5(vf)	
Propyne				0.3(vf)	
Hydrogen cyanide				0.8(vf)	
Propynenitrile				0.07(vf)	
Ethanedinitrile				0.02(vf)	
Oxygen		0.1(ph)	2×10^{-5}		
Sulphur dioxide			2×10^{-4}		
Sodium			2×10^{-8}		

ph, predicted from photolysis but not detected.
sh, from scale height and mean molecular mass.
vf, from simple scaling of predicted downward fluxes and the amount possibly incoporated in the aerosol.

radiation at longer wavelengths than does methane and effectively promotes the photochemical sequence (R9.14)–(R9.18) by generation of a methyl radical:

$$C_2H_2 + h\nu \rightarrow C_2H + H \tag{R9.20}$$

$$C_2H + CH_4 \rightarrow C_2H_2 + CH_3 \tag{R9.21}$$

Haze layers in this cold atmosphere are likely to be derived from products of photolysis. This is probably composed mostly of ethane and ethyne, but with small traces of butadiyne (biacetylene C_4H_2) produced from

$$C_2H_2 + C_2H \rightarrow C_4H_2 + H \tag{R9.22}$$

Over time the photolysis of methane must have led to a very substantial loss of hydrogen from the moon. By contrast the losses of nitrogen have probably been small via impact dissociation (R9.4). The nitrogen atmosphere is thought be reactive enough to allow the formation of a range of nitriles (e.g. hydrogen cyanide, propynenitrile CHCCN and ethanedinitrile, C_2N_2).

Triton
Before the Voyager 2 encounter with Triton, the largest moon of Neptune, it was thought to possess a methane atmosphere. The moon has a temperature perhaps as low as 43 K. This implies a vapour pressure for methane of only 0.1 Pa. At this pressure, methane could be retained by the body for the lifetime of the solar system, and other atmospheric gases candidates, such as nitrogen and argon, would be effectively retained.

Studies of Triton have established the existence of a detectable atmosphere with a pressure of about 1.6 Pa. It consists mostly of nitrogen with 0.02–1% carbon monoxide, but only traces of methane. Extreme seasonal changes in temperature on Triton lead to extensive volatilisation and deposition of ices, but as yet no nitrile ices have been found. We would expect them in much the same way as they are predicted within the atmosphere of Titan. There are some active geysers that emit volatiles from the surface, and there are clouds and haze layers. As with Titan, the aerosol is derived from methane photolysis.

Pluto
There have been no spaceraft missions to Pluto, but two factors have particularly advanced the study of the outermost of the planets in the solar system. It is not always the most distant known body, because its

highly elliptical orbit carries it well within the orbit of Neptune at times. Its recent close approach, and the discovery of a moon, Charon, have improved understanding dramatically. In particular, the alignment of Pluto and Charon has been such in recent years that there have been numerous Pluto–Charon eclipses. Studies of these have revealed much about the planet's surface composition and atmosphere. Observations suggest methane ices at the surface. At 60 K the pressure above methane amounts to 10 Pa, but the ices are probably colder than this. It is reasonable to imagine an atmosphere maintained by sublimation of surface deposits. However, recent measurements of the atmospheric scale height imply an atmosphere with much higher atomic mass (about 25 amu). The current view is that the atmosphere of Pluto is a little like that of the Earth and dominated by nitrogen, but with substantial amounts of carbon monoxide.

9.4   Atmospheres of the outer planets

The planets with substantial atmospheres seem to fall into two distinct compositional types: those planets whose atmospheres resemble that of the sun and the other planets with atmospheres whose compositions do not follow the cosmic abundances closely and are much more variable. A strong resemblance between the composition of the bulk of the material in the solar system and the sun has given rise to the generally accepted theory for the origin of the solar system. It formed from a rotating disc of gas, with a composition similar to that of the sun, some 4.6 Ga ago. Hydrogen and helium are by far the most abundant elements in the solar system. All the other elements made up only a few percent of the solar nebula from which the planets formed. These heavier elements would have been found as a small amount of ice-forming materials such as water, ammonia and methane and even less material of low volatility that formed dust and rocks. Only cold and massive planets can retain substantial quantities of atmospheres dominated by nearly incondensable elements such as hydrogen or helium.

The planets Jupiter, Saturn, Uranus and Neptune have deep atmospheres of hydrogen and helium present at a ratio of about 10:1. This is close to the solar ratio and agrees with the notion that the massive outer planets have accumulated from the solar nebula. Deep within the atmosphere, the hydrides of many other elements are found. Methane, ammonia and water (i.e. the hydrides of carbon, nitrogen and oxygen) are important minor constituents of all their atmospheres.

Although there are large differences in the temperature and masses of these planets, their atmospheres are quite similar. This allows us to discuss the atmosphere of Jupiter in detail and neglect the others except where they vary markedly from that of the larger planet. The planets are composed almost entirely of volatile compounds and possess only a small rocky or metallic core. This accounts for their low density.

If we look at the composition of Jupiter (Fig. 9.4), there is reasonable agreement between solar and Jovian elemental abandances. The wide variety of more exotic hydrides detected in the Jovian atmosphere suggests that minor element abundances do not depart markedly from those of the sun (see Fig. 9.4). Other factors such as their free energy of formation, photochemistry and volatility also govern their presence in the upper atmosphere. A range of hydrides, such as that of tin, have been postulated as constituents within the Jovian atomsphere that should be detectable.

There are further reasons for such agreement between solar and planetary abundance not being perfect. There is an enrichment of

Fig. 9.4. Abundances of some elements in the Jovian atmosphere compared with their solar abundances. Elements with the same relative abundance on Jupiter and the sun would lie along the line. The triangles mark upper limits for undetected hydrides.

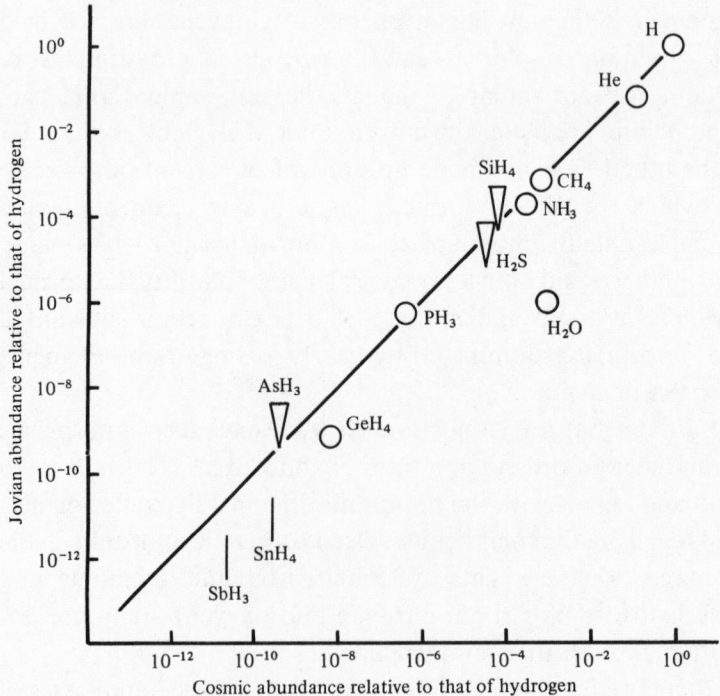

elements heavier than helium on the outer planets. This enrichment increases as the mass of the planet decreases, see Table 9.4. In particular, note how methane, which is accurately known on the outer planets is significantly enriched on Uranus and Neptune. The comparisons are not always easy because on Uranus nitrogen is depleted, but it is thought that this arises through the formation of deep and massive NH_4SH clouds.

Helium in the outer envelope of the giant planets also departs from the solar abundances. This is particularly true for Jupiter, and more especially Saturn. The interior pressures are high enough to allow the formation of metallic hydrogen in the interior. Helium is immiscible with metallic hydrogen, and so differentiates into the core and the envelope of these two planets, and is thus depleted. On Uranus, in contrast, the helium–hydrogen ratio is as expected from solar abundances.

Organic reactions in the Jovian atmosphere
Much modelling work has been done to understand the chemistry of the outer planets so it is possible to give reasonable descriptions of the kind of chemistry that would occur.

Table 9.4. *Some important characteristics of the giant planets*

	Jupiter	Saturn	Uranus	Neptune
Mass (Earth = 1)	320	95	15	17
Radius (km)	71 600	60 000	25 900	24 100
Density (g cm^{-3})	1.31	0.69	1.55	2.23
Tropopause pressure (Pa)	2×10^4	10^4	10^4	10^4
Tropopause temperature (K)	120	85	55?	55?
Enrichment relative to solar abundances				
Oxygen (as H_2O)	1	?	?	?
Carbon (as CH_4)	2.32 ± 0.2	2–6	20–30	25–75
Nitrogen (as NH_3)	2	2–4	?	?
Phosphorus (as PH_3)	$1 + 0.3$	1–8		
Arsenic (as AsH_3)	$1.5^{+1.5}_{-0.8}$	4^{+4}_{-2}		
Helium (as He)	0.7 ± 0.1	0.2 ± 0.1	1 ± 0.2	
Trace atmospheric composition				
Ethane	5	3–6		
Water		1		
Phosphine (PH_3)	0.4	1		
Acetylene	< 2	0.02		
Carbon dioxide	0.1			
Hydrogen cyanide	0.1			
Carbon monoxide	0.002			
Germane (GeH_4)	0.0006			

We have shown that methane is well known in the atmospheres of the outer planets. However, both ethane (5 ppm) and ethyne (i.e. acetylene, at <2 ppm) have been detected together with the methane (650 ppm). The presence of a dehydrogenated compound such as acetylene in the hydrogen-rich Jovian atmosphere is rather surprising. High-energy sources, such as cosmic rays or lightning, are often proposed to maintain their presence. However, there are also less exotic routes to produce unsaturated hydrocarbons, as we saw for the chemistry of Titan, such as ethene and ethyne:

$$CH_4 + h\nu \rightarrow CH_2 + H_2 \qquad \text{(R9.23)}$$

$$CH_2 + CH_3 \rightarrow C_2H_4 + H \qquad \text{(R9.24)}$$

$$C_2H_4 + h\nu \rightarrow C_2H_2 + H_2 \qquad \text{(R9.25)}$$

$$C_2H_4 + H + M \rightarrow C_2H_5 + M \qquad \text{(R9.26)}$$

$$C_2H_5 + H \rightarrow CH_3 + CH_3 \qquad \text{(R9.27)}$$

$$CH_3 + CH_3 + M \rightarrow C_2H_6 + M \qquad \text{(R9.28)}$$

This sequence is probably less efficient on Jupiter than on Titan, because of the high concentration of hydrogen. There have been many proposals that the Jovian atmosphere might support an elaborate organic chemistry. Reactions (R9.20) and (R9.21), in which acetylene acts as a catalyst do not play a major role on Jupiter. Additionally this planet lacks a cool surface on which complex molecules might accumulate and therefore avoid destruction. This discourages high equilibrium concentrations of organic species of greater complexity than ethane. However, there is some spectroscopic evidence for methyl acetylene (propyne, CH_3CCH) and propane on neighbouring Saturn, so one can envisage its production through reactions such as

$$C_2H + CH_3 + M \rightarrow CH_3CCH + M \qquad \text{(R9.29)}$$

which utilise the products of methane photolysis, i.e. (R9.20) and (R9.27).

The atomic hydrogen needed to promote the reaction sequences (R9.23)–(R9.28) can be produced through photolysis at high altitudes:

$$H_2 + h\nu \rightarrow H_2^+ \qquad \text{(R9.30)}$$

$$H_2^+ + H_2 \rightarrow H_3^+ + H \qquad \text{(R9.31)}$$

Unlike that of Titan, the Jovian atmosphere does not lose H, because the gravitational attraction is so large. In addition to this, the time constant for the reaction

$$H + H + M \rightarrow H_2 + M \tag{R9.32}$$

is long enough for the hydrogen atoms to be transported down through the Jovian atmosphere so that they can enter into other reactions. The upper layers of hydrogen act as a shield protecting the methane below the turbopause from photo-ionisation.

The hydrocarbons ethane, ethene and ethyne can readily be converted back to the abundant methane. Other carbon compounds detected in the Jovian atmosphere are carbon monoxide (0.002 ppm) and hydrogen cyanide (0.1 ppm). The latter might be the product of shock synthesis by lightning in the same way as some ethyne.

The production of hydrocarbons from the photolysis of methane is also considered to be an explanation for the fact that the mesospheric temperature of all the giant planets is close to 150 K. This is maintained despite a thirty-fold decrease in solar radiation on going from Jupiter to Neptune. The spectral bands of ethane and ethyne lie at 12.2 and 13.7 μm. These barely overlap with the 150 K black body radiation. If the temperature drops further then the residual overlap vanishes, so the absorption provides a kind of thermostat that can accommodate a wide range of heat inputs.

Chemistry of Jovian nitrogen, sulphur and phosphorus
Hydrogen, helium and methane higher up in the Jovian atmosphere shield the ammonia from UV radiation and limit the rate of photolysis. The most important products of this process are probably hydrazine (N_2H_4), which can condense in the stratosphere, and nitrogen that could be formed from the photolysis of hydrazine:

$$NH_3 + h\nu \rightarrow NH_2 + H \tag{R9.33}$$

$$NH_2 + NH_2 \rightarrow N_2H_4 \tag{R9.34}$$

Small amounts of methylamine (CH_3NH_2) might be expected via a photochemically mediated process:

$$CH_4 + NH_3 \rightarrow CH_3NH_2 + H_2 \tag{R9.35}$$

Hydrogen sulphide is also photolysed by UV radiation that penetrates through breaks in the ammonia clouds:

$$H_2S + h\nu \rightarrow HS + H \tag{R9.36}$$

Subsequent reactions could allow polymerization to S_8 or hydrogen polysulphide chains. It is possible for such sulphur chemistry to produce yellow, orange and brown solids that may be responsible for the markings on the planet and the association of high-temperature areas (greater than 200 K) with brown and yellow clouds. It is also possible for HS to react with ammonia to give rise to the condensable and coloured NH_4SH.

In July and August 1994, many fragments of the comet Shoemaker-Levy collided with Jupiter. There was a dramatic increase in the column abundance of S_2 after impact, which may imply that as much as 10^{14} g was deposited on Jupiter from fragment G. The bimolecular sulphur was probably produced by reactions of hydrogen sulphide with atomic hydrogen:

$$H + H_2S \rightarrow HS + H_2 \tag{R9.37}$$

$$H + HS \rightarrow S + H_2 \tag{R9.38}$$

$$S + S + M \rightarrow S_2 + M \tag{R9.39}$$

The photolysis of phosphine (PH_3) in the Jovian atmosphere is limited by the amount of UV light that can penetrate. However phosphine does not condense in the clouds as ammonia does. Thus, in spite of a much lower abundance than ammonia, at the cloud tops, it becomes more abundant in the stratosphere and protects the ammonia lower down from excessive photodestruction. The principal product of the photolysis of phosphine, red phosphorus, arises as follows:

$$PH_3 + hv \rightarrow PH_2 + H \tag{R9.40}$$

$$PH_2 + PH_2 \rightarrow PH_3 + PH \tag{R9.41}$$

$$PH + PH \rightarrow P_2 + H_2 \tag{R9.42}$$

$$P_2 + P_2 \rightarrow P_4(\text{solid}) \tag{R9.43}$$

Other reaction sequences may lead to the production of phosphoric acid:

$$PH + OH \rightarrow PO + H_2 \tag{R9.44}$$

$$PO + OH \rightarrow PO_2 + H \tag{R9.45}$$

$$PO_2 + PO + M \rightarrow P_2O_3 + M \tag{R9.46}$$

$$2P_2O_3 + M \rightarrow P_4O_6 + M \tag{R9.47}$$

$$P_4O_6 + 6H_2O \rightarrow 4H_3PO_3 \tag{R9.48}$$

Phosphorus chemistry may also protect ammonia in the Jovian atmosphere by reacting as follows:

$$NH_2 + PH_3 \rightarrow NH_3 + PH_2 \tag{R9.49}$$

Jovian clouds

The Jovian planets are dominated by latitudinal bands of cloud that obscure the planetary interior. Upper cloud layers on Jupiter are ammonia-ice. Modelling exercises suggest that below these clouds there will be NH_4SH clouds and yet further down, clouds of water-ice. The altitude and composition of the clouds in relation to the general structure of the atmosphere are given in Fig. 9.5. Theoretical modelling studies usually suggest that the gases of the Jovian troposphere are in a state of chemical equilibrium. The presence of non-equilibrium amounts of

Fig. 9.5. A profile through the atmosphere of Jupiter.

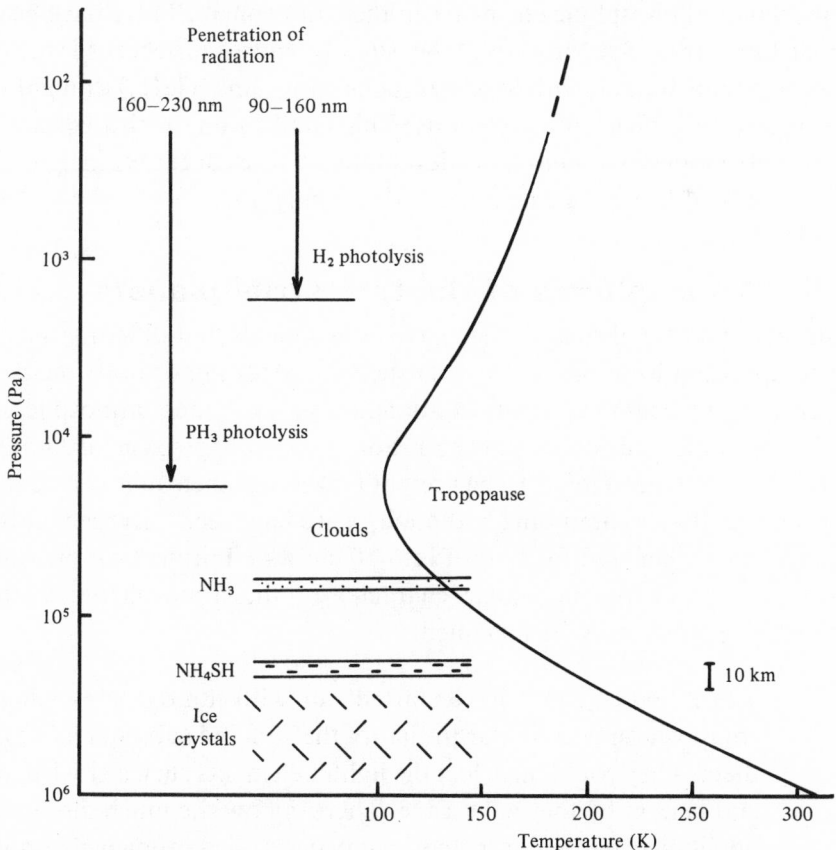

phosphine and carbon monoxide aloft have been taken as an indication of deep atmospheric overturning that extends to the 1000 K level. The high temperatures within the planet are the result of an internal energy source characteristic of the Jovian planets. This energy probably arises from gravitational contraction.

The other giant planets

Our knowledge of the other giant planets, although more restricted, has improved because of several missions. In particular, the Voyager 2 mission to Uranus (1986) and Neptune (1989) represented the first encounter with these remote planetary systems. As with some moons, it appears that haze formation on Neptune involves photolysis of methane, (R9.15)–(R9.18), (R9.20) and (R9.22), which leads to aerosols of ethane, ethyne and butadiyne.

Lower temperatures in Saturn's atmosphere probably mean a more effective removal of ammonia into cloud particles that would increase the importance of phosphine chemistry in the stratosphere. There have been predictions that Saturn should be slightly more enriched in heavy elements than Jupiter, with hydrides such as germane (GeH_4) thought to be present at higher concentrations. This is all in line with nucleation instability models of outer planet development. They argue for increasing enrichment of heavier elements with decreasing size.

9.5   Atmospheres of the terrestrial planets

The inner planets of solar system have atmospheres that differ markedly in composition from that of the solar nebula. Most importantly, neither hydrogen nor helium is the most abundant gases in these atmospheres. Helium is found only at low concentrations and hydrogen is incorporated into water. Nitrogen tends to be present on these planets in its elemental form rather than as ammonia, although, as we have seen, nitrogen is also found, in its elemental form, on Pluto, Titan and Triton. Two possible explanations for the non-solar character of the atmospheres of the terrestrial planets may be imagined.

(i) These bodies could have started out with atmospheres whose composition was similar to that of the sun and subsequently lost them. They could have lost the lighter elements such as hydrogen and helium by thermal escape. There is now the much discussed possibility that major meteoric impacts eroded primaeval atmospheres. Smaller impacts can drive off a fraction of the atmopsphere, but a huge impact could remove the atmosphere entirely.

(ii) Alternatively, the planets may never have captured lighter volatile material from the solar nebula, in which case the atmosphere found today is the result of entirely different secondary processes.

The terrestrial planetary atmospheres are characterised by the presence of significant amounts of oxidised compounds such as water and carbon dioxide (Table 9.5). This is in striking contrast to the dominance of reduced compounds such as hydrides associated with the Jovian planets. The major atmospheric components of the Earth differ from those of Mars and Venus in two respects: (i) Venus and Mars have atmospheres dominated by carbon dioxide while nitrogen is the most abundant component of the atmosphere of the Earth and (ii) the Earth's atmosphere is unique because of the presence of large concentrations of oxygen. As the Earth has been discussed in considerable detail throughout this book, the remainder of this section will concentrate on the composition and chemistry of the atmospheres of Mars and Venus. We will return to the Earth in the final sections of this chapter.

Venus

Venus is covered with thick clouds. It has long been thought that the carbon dioxide in the Cytherean atmosphere (the term Cytherean is

Table 9.5. *The composition of the atmosphere of the inner terrestrial planets. Concentrations beneath that of water in the table are given as ppm*

	Venus	Earth	Mars
Carbon dioxide	96%	0.03%	95%
Nitrogen	3.5%	77%	2.7%
Oxygen	<0.001%	21%	0.13%
Water vapour	≤0.5%	0.01%	0.03%
Helium	10	5.24	<100
Argon	70	9340	16 000
Argon-36	35	31	5
Neon	5–13	18	2.5
Krypton	0.5	1.1	0.3
Xenon	<0.04	0.08	0.08
Carbon monoxide	50	0.1	700
Sulphur dioxide	150	0.01–0.1	
Hydrogen chloride	1	0.001	
Hydrogen fluoride	0.02		
D/H ratio	0.022	0.00015	0.0009

sometimes used in preference to Venusian) could be buffered by chemical interactions with the surface. This is a chemical equivalent to the physical equilibria that have been postulated between surface methane deposits and methane atmospheres on Pluto and Triton. A likely buffer is

$$CaSiO_3 + CO_2 \rightarrow CaCO_3 + SiO_2 \tag{R9.50}$$

The surface temperature and pressure are very high (about 750 K and 9 MPa). Atmospheric gases, such as carbon dioxide should be in equilibrium with the surface rocks. However this is not true of the small amount of nitrogen present in the Cytherean atmosphere that would probably have no effective surface sink.

The most interesting element in the atmosphere of Venus is sulphur. The concentrations of COS, S_2 and H_2 are compatible with an equilibrium with surface minerals, but SO_2 and O_2 concentrations are too large for them to be in equilibrium with the surface. It is probable that there are two sulphur cycles in the Cytherean atmosphere: a fast atmospheric cycle in the upper atmosphere and a slower one at low levels. The fast cycle is represented by

$$2CO_2 \rightarrow 2CO + O_2 \tag{R9.51}$$

$$O_2 + 2SO_2 \rightarrow 2SO_3 \tag{R9.52}$$

$$SO_3 + H_2O \rightarrow H_2SO_4 \tag{R9.53}$$

This process is reversed in the lower atmosphere:

$$H_2SO_4 \rightarrow H_2O + SO_3 \tag{R9.54}$$

$$SO_3 + CO \rightarrow SO_2 + CO_2 \tag{R9.55}$$

These fast reactions are responsible for generating the sulphuric acid component of the clouds and the regeneration of sulphur dioxide at lower altitudes. The slow cycle involves the destruction of carbonyl sulphide at high level by processes such as

$$3O_2 + 2COS \rightarrow 2SO_3 + 2CO \tag{R9.56}$$

with regeneration occuring lower in the atmosphere:

$$CO + S \rightarrow COS \tag{R9.57}$$

Essentially, the processes in the upper atmosphere are photochemically driven and produce oxidised compounds that are thermodynamically unstable in the lower atmosphere.

The chemistry of hydrogen chloride in the Cytherean atmosphere is not so clear. This trace gas is usually assumed to be the dominant source of hydrogen atoms through the reaction

$$HCl + hv \rightarrow H + Cl \tag{R9.58}$$

This reaction provides a source of hydrogen atoms that escape from the atmosphere at a rate of 10^4–$10^7 \, cm^{-2} \, s^{-1}$. The upward transport of hydrogen must be balanced by downward movement of chlorine. One can imagine hydrogen chloride being reformed in the high temperature lower atmosphere through a series of reactions involving interaction with the surface, e.g.

$$CO_2 + 3FeMgSiO_4 \rightarrow CO + 3MgSiO_3 + Fe_3O_4 \tag{R9.59}$$

$$CO + H_2O \rightarrow CO_2 + H_2 \tag{R9.60}$$

$$H_2 + Cl_2 \rightarrow 2HCl \tag{R9.61}$$

It is important to note how this reaction effectively oxidises iron at the surface of the planet using the oxygen in water and losing hydrogen from the upper atmosphere. This is aided by vertical transport as hydrogen chloride followed by photolysis. Chemical reactions at the surface of Venus control much of the chemistry of the atmosphere.

There is convincing evidence that the clouds that so dominate our picture of Venus are composed of a concentrated aqueous solution of sulphuric acid. The vertical structure of the Cytherean atmosphere is shown in Fig. 9.6, together with the position of various cloud layers. The smallest cloud particles could be elemental sulphur, while the larger droplets found in the lower cloud layers are probably sulphuric acid. These clouds contain little water and remind us how depleted the planet is in water. The absence of water is the subject of much debate. Did Venus lose its original inventory of water or did it never have any water in the first place?

Mars

The Martian atmospheric composition is well known from many observations from space probes and from equipment soft-landed on the surface. The atmosphere is predominantly carbon dioxide. Its surface pressure shows a pronounced seasonal cycle as the carbon dioxide is deposited and re-sublimated at the south polar cap.

The photochemistry of the Martian atmosphere is dominated by the photolysis of carbon dioxide:

$$CO_2 + hv \rightarrow CO + O \tag{R9.62}$$

This being so, it is a little puzzling, at first, that the carbon dioxide atmosphere prevails over one of oxygen or carbon monoxide. However, oxygen is probably produced by the reaction

$$O + HO_2 \rightarrow OH + O_2 \tag{R9.63}$$

This reaction also provides a source of hydroxyl radicals that can rapidly oxidise carbon monoxide:

Fig. 9.6. The temperature–pressure relationship and cloud composition for Venus.

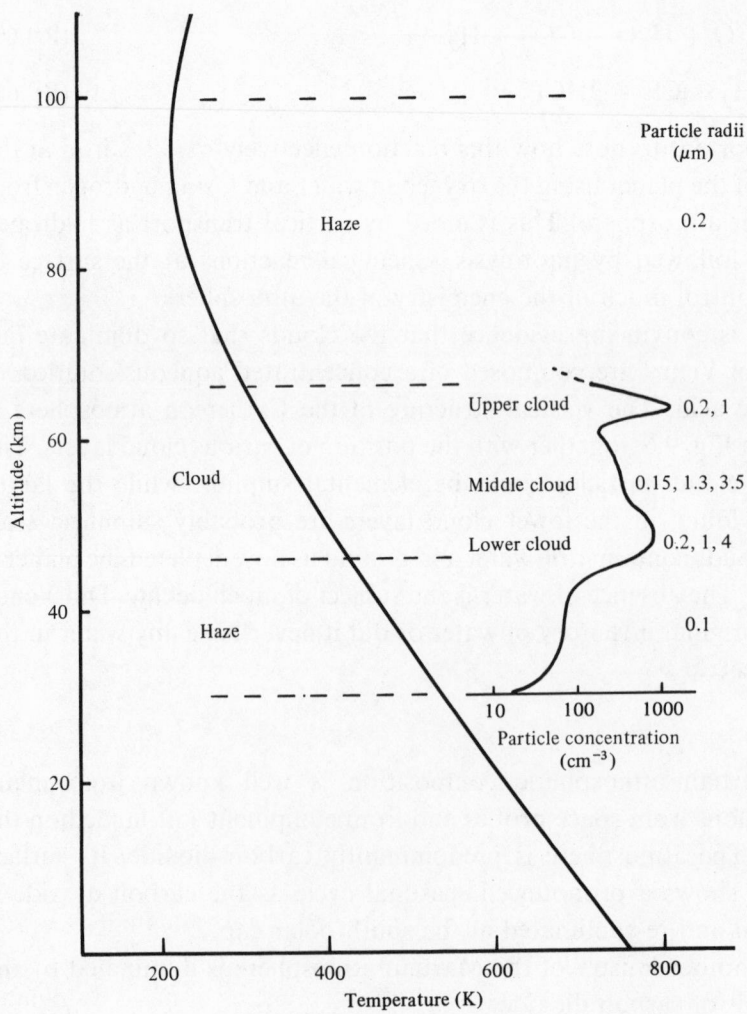

$$OH + CO \rightarrow CO_2 + H \tag{R9.64}$$

The hydrogen atom can be involved in some subsequent reactions, one of which will generate HO_2 with carbon dioxide acting as a third body and remove oxygen from the atmosphere:

$$H + O_2 + CO_2 \rightarrow HO_2 + CO_2 \tag{R9.65}$$

Measurements suggest that the concentration of carbon monoxide is too high for this picture to be entirely correct. Possibly HO_2 concentrations are lowered by the presence of aerosols in the Martian atmosphere that absorb HO_2.

The question of liquid water on Mars has been as long debated as the question of water on Venus. Mars has many land-forms that seem reminiscent of erosion features on the Earth, yet the temperatures on Mars are currently too low to allow the flow of water. However, early Mars may have been warmed by a greenhouse effect sufficiently to melt water-ice. If large quantities of water were placed in the Martian atmosphere, then OH and H concentrations would increase through photochemical reactions. This would allow high concentrations of oxygen to exist on ancient Mars. The oxygen increases could be long-lived, because the excess oxidising power of the atmosphere could only be removed by two slow processes: (i) escape to space or (ii) reaction with Fe(II) at the planetary surface.

Oxygen, even in its atomic form, is quite heavy and would escape only slowly from the Martian exosphere. Therefore, we have to imagine another mechanism for escape. This is similar to the photo-ionisation described for the loss of gases from the lunar exosphere, i.e.

$$O + h\nu \rightarrow O^+ + e^- \tag{R9.66}$$

but other loss mechanisms are possible:

$$CO^+ + e^- \rightarrow C + O \tag{R9.67}$$

$$O_2^+ + e^- \rightarrow O + O \tag{R9.68}$$

$$O_2 + h\nu \rightarrow O + O \tag{R9.69}$$

The fragments produced in these reactions can have enough energy to escape from the upper atmosphere.

The two *Viking* landers carried mass spectrometers that operated during the descent onto the planet. This means that we have a good picture of the composition of the atmosphere which remains dominated by carbon dioxide up to at least 180 km. In the upper atmosphere (120–180 km), electrons are mainly produced through photo-ionisation of carbon dioxide:

$$CO_2 + h\nu \rightarrow CO_2^+ + e^-$$ (R9.70)

CO_2^+ is removed by

$$CO_2^+ + e^- \rightarrow O + CO$$ (R9.71)

$$CO_2^+ + O \rightarrow O_2^+ + CO$$ (R9.72)

$$CO_2^+ + O \rightarrow O^+ + CO_2$$ (R9.73)

these reactions ensure that O_2^+ is the dominant ion below 300 km altitude. The O^+ ion is not a major component of the Martian ionosphere, unlike the terrestrial situation, because its reaction with plentiful carbon dioxide is rapid:

$$O^+ + CO_2 \rightarrow O_2^+ + CO$$ (R9.74)

9.6   The origin and evolution of secondary atmospheres

The previous section drew attention to the striking differences between the atmospheres of the inner planets and those found elsewhere in the solar system. The composition of the Jovian planets is similar to the original solar nebula. The inner planetary atmospheres are so different that the origin of their atmospheres requires different processes. There are many problems in our understanding of the evolution of the Earth's atmosphere, so the discussion remains tentative.

The primordial atmosphere

Table 9.6 compares the terrestrial abundance of elements to that of the sun. Expressing these abundances relative to silicon (a non-volatile element), we see that the Earth is deficient in the light volatile elements and the inert gases. During their formation, the giant planets were sufficiently distant from the sun to be cool enough to condense the many more volatile elements of the solar nebula or to capture the nebula virtually with cosmic abundances. Of course, once massive planets had formed in the inner solar system, it might have been possible for them to capture the more volatile compounds as atmospheres. Such atmospheres could have resembled the nebula in composition and had large quantities of the abundant hydrogen and helium. However, we can see that today, even if we include the hydrogen present as water in the hydrosphere and atmosphere of the Earth, the planet is found deficient in hydrogen. Helium has been depleted to an even greater extent. A popular theoretical model, to explain this, sees the planets as condensing from the solar nebula, with only the refractory materials solidifying close to the sun. The more volatile components will condense only on planets at greater distances from the sun. This relatively simple view would lead to the argument that the inner planets are so close to the sun that they could simply never accumulate these volatile materials. It is an attractive theory, but it does not satisfactorily explain all of the features that we see in the terrestrial atmospheres.

Another approach would be to argue that the terrestrial planets accumulated atmospheres that they subsequently lost. If we were to consider a thermal mechanism, it would be only the lightest elemenst that would be lost, so we would expect to see atmospheres that were depleted in H and He. Consider a heavy inert gas such as neon. This should not have escaped over the lifetime of the Earth, so it is possible to calculate the amount of the other elements present together with neon, in a hypothetical primordial atmosphere. We do this by using the solar abundance given in Table 9.6. There is currently 6.48×10^{16} g of neon in the Earth's atmosphere. This implies 3.3×10^{19} g of hydrogen, far less than the 1.7×10^{23} g found as water on the Earth. Thus even if the Earth retained

Table 9.6. *Solar and terrestrial abundances of the elements relative to silicon* $(Si = 10^6)$.

Data are from Mason (1966) and Cameron (1980).

Element	Abundance	
	Cosmic	Crustal
Hydrogen	2.66×10^{10}	8400
Helium	1.80×10^{9}	3.5×10^{-5}
Oxygen	1.84×10^{7}	3.5×10^{6}
Carbon	1.11×10^{7}	7100
Neon	2.60×10^{6}	1.2×10^{-4}
Nitrogen	2.31×10^{6}	21
Magnesium	1.06×10^{6}	8.9×10^{5}
Sulphur	5.00×10^{5}	3.5×10^{5}
Argon	1.06×10^{5}	0.06
Sodium	6.0×10^{4}	4.6×10^{4}
Krypton	41.3	6×10^{-6}
Xenon	5.84	5×10^{-7}

some of the primordial atmosphere it captured from the solar nebula, this can obviously account for only a small fraction of the atmosphere that is present today.

It may have been that the Earth had a very extensive primordial atmosphere, but that this was entirely lost at an early stage by a mechanism such as impact erosion. In this process, the energy of large meteorite (bolide) impacts would have torn away substantial fractions of the atmosphere. The most significant event, with this regard, in Earth's history was the collision of a Mars-sized body with the Earth that created the moon. An impact of this magnitude would have eroded the entire atmosphere.

It is also possible for the heavy molecules of an atmosphere to be blown of by a stream of lighter hydrogen. When the sun was more active in early solar system history (the *T Tauri* phase), UV intensities would have been much higher. This could have caused massive dissociation of atmospheric water vapour that could carry heavier molecules with it and offer a further model for the loss of a primaeval atmosphere.

The estimate of the mass of the primordial atmosphere made above is an upper limit. It is likely that much of our neon inventory was not acquired as a gaseous envelope, but arrived absorbed on solids that were captured by the early Earth. Evidence that the rare gases arose this way and not from post-accretional capture of gases of the solar nebula can also be obtained from measurements. The rare gas composition on the Earth more closely resembles that found in meteorites than it does that found in solar material (Fig. 9.7). Xenon is under-abundant on the Earth relative to argon and krypton, which fact may be due to the selective adsorption of xenon onto shales.

Models for the origin of the solar system involving condensation or capture of the original solar nebula are still attractive. However, it is important to look at secondary sources for atmospheres of the inner planets. Many theories of formation of planets suggest an 'inhomogeneous accretion' of material. Such theories argue that the planets initially formed from volatile-poor rocky material resembling stony meteorites. Later, as the solar nebula cooled, volatile-rich meteorites formed. The subsequent accretion of these meant that planets acquired a coating of volatile elements. There are some meteoritic analogues of both the volatile-rich and volatile-poor planet-forming materials. However, none satisfy all the constraints for constructing the planets of the solar system from one or two types of meteoritic material.

In a loose way, both the meteoritic and the equilibrium-condensation models seem to satisfy the observation that Mercury lacks volatile material while large amounts are found on the outer planets. The difficulties in chosing between them are not to be underestimated. The interested reader will find the subject well reviewed in the additional reading.

Out-gassing

No matter which method of accretion we adopt to explain the formation of the planets, most theories suggest that, at some early date, the outermost layers of the terrestrial planets were volatile-rich. These volatiles could subsequently be released by heating in a process known as out-gassing. There are two obvious sources for this heat: radiogenic and possibly accretional heating, for a short time at least, from the impact of large meteorites. The radiogenic heat source could well have come from short-lived radio-isotopes such as ^{129}I and ^{244}Pu and over a longer period

Fig. 9.7. Terrestrial and meteorite abundances compared with those of the sun (atoms per 10^6 atoms of silicon). The diagonal line shows the position of terrestrial elements with their solar abundances. (From Mason (1966) and Cameron (1980).

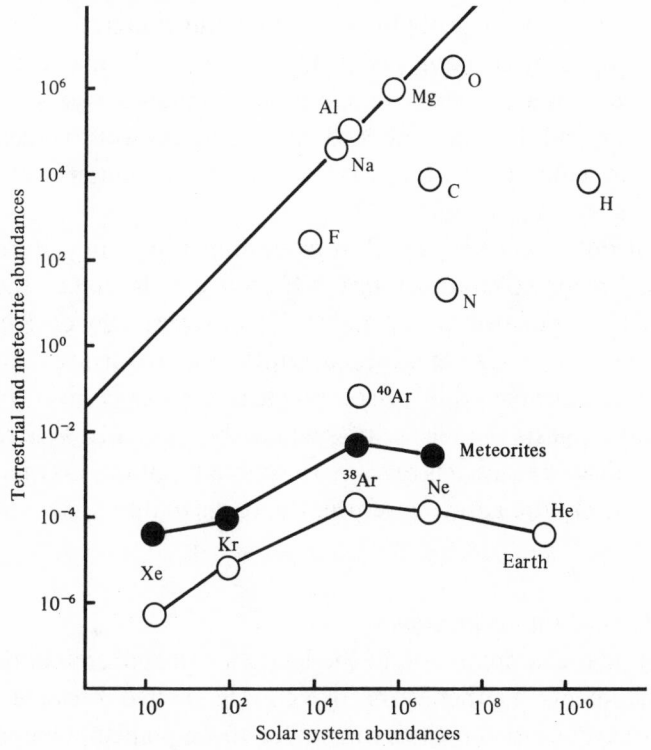

from ^{40}K, ^{232}Th, ^{235}U and ^{238}U. Heat from this radioactive decay would drive volatile materials into the atmosphere of a young planet.

The difficulty in trying to establish the early composition of the Earth's atmosphere is hampered by the absence of sedimentary rocks older than 3.8 Ga. For many years a strongly reducing atmosphere rich in methane and ammonia was thought probable. This was proposed for a number of reasons: geological evidence suggests that the Earth's surface was more reducing in the past and methane and ammonia are common elsewhere in the solar system. It is also implied by the origin of life on the Earth that probably took place after most of the volatiles had been de-gassed. This latter point was a most powerful argument for a highly reducing early atmosphere on the Earth, because it appeared that the molecular precursors to life could only be stabilised on a highly reducing Earth. However, increasing knowledge of the photochemistry of planetary atmospheres suggests that methane–ammonia atmospheres have only limited stability. If such an atmosphere had existed on the early Earth, it would have been rapidly destroyed through photolysis. This would be followed by thermal escape of hydrogen, to leave an atmosphere of nitrogen and carbon dioxide. Current views of the Earth's earliest atmosphere see it as dominated by high carbon dioxide partial pressures. Nitrogen and water vapour were probably important additional constituents. Vast amounts of water accumulated in the oceans, and the carbon dioxide was dissolved in precipitated sedimentary carbonates. Gases such as hydrogen sulphide, sulphur dioxide and hydrogen chloride vented during the de-gassing would have rapidly reacted with solid materials to contribute sulphates to the oceans.

Some support for out-gassing is gained by examining the composition of volatiles released from volcanoes today. Of course, this makes the assumption that the composition of the mantle, from which they derive, has not changed very much with time. It is currently more oxidising than it was in the past. The composition of some volcanic gases is given in Table 9.7, but it still suggests the release of much carbon dioxide. Where oxygen is found in volcanic gases, it is regarded as contamination. Oxygen was probably not among the gases present in the earliest atmosphere of the Earth.

The volatile inventory of the inner planets
We can at least know the composition of the Earth's volcanic emissions from current measurements. Neither Mars nor Venus has active volcanoes, so there is no possibility of such measurements for those planets. One of

the most difficult problems concerning the origin of the atmospheres is the great difference between the atmospheres found on these inner planets. Table 9.8 shows the current volatile inventories of the terrestrial planets. For Venus and Mars these represent minimum values because our knowledge of their crustal composition is poor.

Venus and the Earth seem to have similar quantities of carbon dioxide, if we consider all the Cytherean carbon dioxide to be present in the atmosphere of Venus. Nitrogen is also present in roughly equivalent amounts because it does not interact with the surface. However, the striking difference is that very little water is present on Venus. Either none accreted with the planet or an enormous amount of water was subsequently lost from the Cytherean atmosphere. Photodissociation of water followed by escape of hydrogen is the mechanism frequently invoked to explain why Venus is so dry. The recent observation that Venus has a high D/H ratio (enriched by a factor of 150 above that of the Earth) suggests that the planet has lost a great deal of water (see Table 9.6). High D/H ratios arise

Table 9.7. *The composition of volcanic gases from Kilauea, Hawaii.*

Data are from Heald, Naughton and Barnes (1963) and Gerlach and Graeber (1985).

	Range	East Rift mean
Water	Major component	80.3
Hydrogen	0–0.43%	0.92
Carbon dioxide	0.00023–3.8%	3.67
Carbon monoxide	0–0.086%	0.066
Methane	0–0.047	
Hydrogen sulphide	0–0.12%	1.02
Sulphur dioxide	0–0.16	13.6
Hydrogen chloride		0.16
Hydrogen fluoride		0.18

Table 9.8. *The curent volatile inventory for the inner planets (g).*

Values for the Earth include the atmosphere, the hydrosphere and the sedimentary shell. Values for Venus include only the atmosphere. Values for Mars include the atmosphere and the polar cap.

	Earth	Venus	Mars
Carbon dioxide	2.5×10^{23}	5.3×10^{23}	10^{20}
Water	1.6×10^{24}	$<1.9 \times 10^{21}$	
Nitrogen	4.9×10^{21}	6.1×10^{21}	2×10^{17}
Chlorine	3.06×10^{22}	4×10^{17}	

because the loss of water via hydrogen escape from the exosphere will proceed much faster for normal hydrogen than for the heavier deuterium atom. However, this hypothesis requires the production of an enormous amount of oxygen that would have oxidised minerals at the surface. Alternatively, we could argue that the water was lost in the *T Tauri* phase. At this time, water would have been rapidly photolysed and many of the heavier oxygen atoms blown off from the atmosphere with the out-flowing hydrogen. However, this mechanism would require rapid photolysis of oceans of water.

We are left with similar uncertainties when considering the volatile inventory of Mars. Mars is very deficient in volatiles, if one counts only those present in the atmosphere and the polar caps. There are three ways in which this could be explained: (i) the planet never accumulated much volatile material, (ii) it had a large volatile inventory, but much of this escaped or (iii) the planet never out-gassed as actively as the Earth, or if it did, then the volatiles are currently buried on the planet. In particular, it would be useful to know whether there had ever been enough carbon dioxide (10^5 Pa) to warm Mars by a greenhouse effect and allow substantial volumes of liquid water to exist. There is insufficient carbon dioxide in the polar cap to achieve this, but many theorists have argued that it is locked up as carbonate minerals on the surfaces. Unfortunately the spectroscopic evidence for the presence of these carbonates is still indefinite. Furthermore, it is likely that a carbon dioxide greenhouse effect would be insufficient to raise the Martian temperature above freezing point because of the lower luminosity of the early sun. It might be necessary to postulate the addition of other greenhouse gases such as ammonia or sulphur dioxide to establish temperatures high enough to allow water to flow.

There is currently only a small amount of nitrogen in the volatile inventory of Mars. However the lack of nitrogen cannot be used as an indicator that little out-gassing occurred, because nitrogen could easily have escaped from the atmosphere through the dissociative mechanisms

$$N_2 + h\nu \rightarrow N_2^+ + e^- \tag{R9.75}$$

$$N_2^+ + e^- \rightarrow N + N \tag{R9.76}$$

Mars could have lost as much as 10^5 Pa of nitrogen in such a way throughout the age of the solar system. The high ratio of $^{15}N : ^{14}N$ in the Martian atmosphere favours a pressure of several hundred pascals of nitrogen at an early date. There is geological evidence for massive deposits of ice (water and perhaps carbon dioxide), which lends weight to the idea of the burial of large amounts of volatile material.

9.7 Life and the evolution of our atmosphere

In Chapter 1 we noted the remarkable similarity between the elemental composition of the atmosphere of the Earth and that of living material. This chapter emphasised the uniqueness of the Earth's atmosphere. It is very far from chemical equilibrium, which fact is a strong indicator of active biological processes.

The earliest atmosphere

It is now believed that the Earth accreted quite rapidly on a geological time scale. This probably took only 10–100 Ma. Atmospheric loss by impact erosion or *T Tauri* blow-off would have greatly depleted any primordially acquired atmosphere. After accretion had finished, the loss of heavier gases would have become much more difficult. Once the planet had formed, metallic iron would have been rapidly removed to the core. This sequestered enormous amounts of this strongly reducing element from the upper parts of the planet, allowing the upper mantle to become more oxidising.

Earlier theories of the origin of the atmosphere, which favoured strongly reducing conditions with abundant methane and ammonia, find less support now. The post-accretional atmosphere was probably 10^6 Pa in carbon dioxide and monoxide and perhaps 10^5 Pa in nitrogen. The carbon dioxide can readily be removed as carbonates on an ocean-covered Earth, so, just before the origin of life, it is likely that the surface atmosphere was

nitrogen	10^5 Pa
carbon dioxide	10^4 Pa
hydrogen	10^2 Pa
oxygen	10^{-8} Pa

Although molecular oxygen would have been at very low concentrations at the surface, photolysis could have maintained it at 1% or so high up in the atmosphere.

Precursors to life

These moderately oxidising conditions do not particularly favour formation of organic molecules, which are necessary precursors to life. In the classic experiments of Urey and Miller, amino acids and many other organic molecules were generated when electrical discharges were put through ammonia–methane atmospheres. However, similar experiments are not particularly successful in more oxidised atmospheres. Although it has to be admitted that surprisingly small amounts of ammonia can stabilise amino acids. Of course, one possible source for the organic molecules is interstellar clouds. These have an elaborate organic chemistry and molecules as complex as ethanol (CH_3CH_2OH) and propionitrile (CH_3CH_2CN) have been detected. These are generated in reactions that arise through the interaction of cosmic rays and gas in interstellar clouds. Alternatively, the large set of organic compounds that are to be found in the heads of comets might be seen as a source of precursor molecules. However, terrestrial sources are favoured for early organic material.

Precursor molecules are much more difficult to produce in carbon dioxide atmospheres. The photochemistry of the mildly reducing atmosphere would yeild abundant HCHO. However, nitrogen fixation is required to form the pre-biotic compounds. Hydrogen cyanide would be a likely candidate molecule, but it is hard to form because of the need to break the triple bond of molecular nitrogen. It can be done with lightning shock fronts, but it is not clear that the recombining fragments would form HCN rather than a nitrogen oxide. If HCN and HCHO dissolved in the early ocean they could react to form more biologically interesting compounds such as the amino acid serine:

$$HCHO + HCN + H_2O \rightarrow NH_2CH \underset{\diagdown COOH}{\overset{\diagup CH_2OH}{}} \tag{R9.77}$$

Alternatively, ferrous iron in the early ocean might have provided reducing power to aid early synthetic reactions. This metal ion can catalyse production of aldehydes under UV irradiation and allow carbon dioxide/water atmospheres to yield glycoaldehyde and ultimately sugars:

$$2HCHO \rightarrow HOCH_2CHO \tag{R9.78}$$

The surfaces of clays offer a further possibility, because they could provide templates for the production of organic molecules.

Early life
Despite the difficulties of arriving at a satisfying picture of these very early steps, it is assumed that life arose on an Earth where organic molecules

produced by abiotic processes were quite abundant. The earliest organisms probably made use of the energy incorporated in these molecules by the simple but inefficient process of fermentation. Of course, the process is still in evidence among yeast today and can be represented by

$$C_6H_{12}O_6 \rightarrow 2CH_3CH_2OH + 2CO_2 \tag{R9.79}$$

or the reaction:

$$3C_6H_{12}O_6 + H_2O \rightarrow CH_3CHOHCOOH + CH_3COOH + CO_2 + 2C_6H_{14}O_6$$
$$\text{fructose} \qquad\qquad \text{lactic acid} \qquad\qquad \text{acetic acid} \qquad\qquad \text{mannitol}$$
$$\tag{R9.80}$$

found in some *Lactobaccilli*. These reactions generate carbon dioxide that would have been released to the early atmosphere, but there was already an enormous amount present so this would hardly have had an important impact. The fermentation would gradually have consumed the readily available organic material.

Autotrophic life forms signalled a major advance. These could synthesise their own organic material using other reactions as a source of energy, e.g.

$$2CO_2 + 4H_2 \rightarrow CH_3COOH + 2H_2O \tag{R9.81}$$

or

$$CO_2 + 4H_2 \rightarrow CH_4 + 2H_2O \tag{R9.82}$$

Alternatively, autotrophs might have developed a photosynthetic mechanism using hydrogen sulphide as a source of hydrogen and precipitating metallic sulphides (e.g. FeS_2). Water could have been another source of hydrogen, but at this stage in the evolution of life it is unlikely that oxygen was released. More probably it would have been used to oxidise ferrous iron.

The really important impact of autotrophs on the atmosphere of the Earth occurred about 2 Ga ago with the appearance of oxygen-producing photosynthetic organisms. These organisms could use water rather than hydrogen as an electron donor. This would have involved solving various biological problems associated with the toxicity of oxygen and the reactivity of several radical intermediates. Nevertheless, protection systems evolved and gradually these micro-organisms became very successful, freed from the limited sources of reduced compounds in the environment.

The release of oxygen into the atmosphere can be traced, in the sedimentary record, by the appearance of large qualtities of oxidised iron in sedimentary rocks. It is not clear why it rose to high concentrations so rapidly about 2.0 Ga ago. Some have argued that it was rapidly produced by photosynthetic organisms. Others suggest that the output of volcanogenic hydrogen decreased and that this lowered the reducing ability of the atmosphere.

The presence of oxygen had an enormous impact. Small amounts, generated by the photolysis of water, had been tolerated for a long time, but the concentrations became so large that it was a major component of the atmosphere. Organisms on the Earth had to adapt: by burying themselves in anoxic regions or developing a tolerance to the gas. The presence of oxygen was not entirely negative because it gave rise a new metabolic opportunity, the very efficient process of respiration.

Thus the atmosphere of the Earth grew into the unique system we see today.

9.8   Further reading

Croswell, K. (1991) A moon with atmosphere, *New Scientist*, **131**(1788), 41–4.

Croswell, K. (1992) Nitrogen stays fixed in pluto atmosphere, *New Scientist*, **134**(1826), 19.

Hunten, D.M. (1993) Atmospheric evolution of the terrestrial planets, *Science*, **259**, 915–20.

Kasting, J.F. (1993) Earth's early atmosphere, *Science*, **259**, 920–26.

Lewis, J.S. and Prinn, R.G. (1984) *Planets and their Atmospheres*, Academic Press, Orlando, Florida.

Lunine, J.I. (1989) Origin and evolution of outer solar system, *Science*, **245**, 141–7.

Squyres, S.W. and Kasting, J.F. (1994) Early Mars – how warm and how wet? *Science*, **265**, 744–9.

Taylor, F.W. (1994) The atmospheres of the inner planets, *Current Science*, **66**, 512–24.

Toon, O.B. (1992) A physical model of Titan's aerosols, *Icarus*, **95**, 24–53.

REFERENCES

Chapter 1

Arrhénius, G. O. S. (1959) In *Researches in Geochemistry*, ed P. H. Abelson, John Wiley, New York.

Hunter-Smith, R. J., Balls, P. W. and Liss, P. S. (1983) Henry's law constant and the air–sea exchange of various low molecular weight halocarbon gases, *Tellus*, **35B**, 170–6.

Junge, C. E. (1963) *Air Chemistry and Radioactivity*, Academic Press, New York.

Junge, C. E. and Seiler, W. quoted in Newell, R. E. (1971) The global circulation of atmospheric pollutants, *Scientific American*, Jan.

Telegardas, K. and Ferber, G. J. (1975) Atmospheric concentrations and inventory of krypton-85 in 1973, *Science*, **190**, 882–3.

Chapter 2

Crutzen, P. J. and Andreae, M. O.(1990) Biomass burning in the tropics, *Science*, **250**, 1669–78.

Nriagu, J. O. (1989) A global assessment of natural sources of atmospheric trace metals, *Nature*, **338**, 47–9.

Jaenicke, R. (1982). Physical aspects of the atmospheric aerosol. In *Chemistry of the Polluted and Unpolluted Troposphere*, ed. H. W. Georgii and W. Jaeschke, pp. 341–67, D. Reidel, Dordrecht.

Chapter 4

Baldwin, A. C. (1982) Reactions of gases on prototype surfaces. In *Heterogeneous Atmospheric Chemistry*, ed. D. R. Schryer, pp. 92–102, American Geophysical Union, Washington DC.

Blanchard, D. C. (1983) The production, distribution and bacterial enrichment of the sea-salt aerosol. In *Air Sea Exchange of Gases and Particles*, ed P. S.Liss and W. G. N. Slinn, pp. 407–454. D. Reidel, Dordrecht.

Hesketh, H. E. (1977) *Fine Particles in Gaseous Media*, Ann Arbor Science Publishers, Ann Arbor.

Hidy, G. M. (1973) Removal processes of gaseous and particulate pollutants. In *Chemistry of the Lower Atmosphere*, ed. R. I. Rasool, pp. 121–76, Plenum Press, New York.

Jaenicke, R. (1978) *Berichte der Bunsengesellschaft für physikalische Chemie*, **82**, 1198–202.

June, C. E. (1963) *Air Chemistry and Radioactivity*, Academic Press, New York.

Nriagu, J. O. (1989) A global assessment of natural sources of atmospheric trace metals, *Nature*, **338**, 47–9.

Penkett, S. A., Atkins D. H. F. and Unsworth, M. H. (1979) Chemical composition of the ambient aerosol in the Sudan Gezira, *Tellus*, **31**, 295–307.

Ponche, J. L., George, C., Mirabel, P. (1993) Mass-transfer at the air–water-interface – mass accommodation coefficients of SO_2, HNO_3, NO_2 and NH_3, *Journal of Atmospheric Chemistry*, **16**, 1–21.

Sinclair, D. (1950) *Handbook of Aerosols*, USA Atomic Energy Authority, Washington DC.

Slinn, W. G. N. (1983) Air-to-sea transfer of particles. In *Air Sea Exchange of Gases and Particles*, ed P. S. Liss and W. G. N. Slinn, pp. 299–407, D. Reidel, Dordrecht.

Tang, I. N., Wong, W. T. and Munkelwitz, H. R. (1981) The relative importance of atmospheric sulphates and nitrates in visibility reduction. *Atmospheric Environment*, **15**, 2463–71.

Warneck, P. (1988) *Chemistry of the Natural Atmosphere*, Academic Press, San Diego.

Willeke, K. and Whitby, K. T. (1975) Atmospheric aerosols: size distribution interpretation, *Journal of the Air Pollution Control Association*, **25**, 529–34.

Chapter 5

Bormann, B. T., Tarrant, R. F., McMclellan, M. H., Savage, T (1989) Chemistry of rainwater and cloud water at remote sites in Alaska and Oregon, *Journal of Environmental Quality*, **18**, 149–52.

Camarero, L. and Catalan, J. (1993) Chemistry of bulk precipitation in the central and eastern pyrenees, northeast Spain, *Atmospheric Environment*, **27**, 83–94.

Seinfeld, J. H. (1986) *Air Pollution*, John Wiley and Sons, New York.

Chapter 6

Bravo, H. A., Guadalupe, R.-O. R., Sosa, R. E. and Torres, R. J. (1991) Report of the historical trends of the levels of ozone monitored at the suburban monitoring station of the University of Mexico at Mexico City, Air and Waste Management Association Preprint 91–115.2.

Campbell, I. M. (1977) *Energy and the Atmosphere*, John Wiley and Sons, London.

Leighton, P. A. (1961) *Photochemistry of Air Pollution*, Academic Press, New York.

Perkins, H. C. (1974) *Air Pollution*, McGraw-Hill Kogakusha, Tokyo.

Poissant, L., Schmit, J. P. and Beron, P. (1994) Trace inorganic elements in rainfall in the Montreal Island, *Atmospheric Environment*, **28**, 339–46.

Sakugawa, H., Kaplan, I. R., and Shepard, L. S. (1993) Measurements of H_2O_2, aldehydes and organic-acids in Los Angeles rainwater – their sources and deposition rates, *Atmospheric Environment* B, **27**, 203–19.

Smirnioudi, V. N. and Siskos, P. A. (1992) Chemical-composition of wet and dust deposition in Athens, Greece, *Atmospheric Environment* B, **26**, 483–90.

Chapter 7

Derwent, D. and McMullen, T. (1993), *Local Air Quality Management*, NSCA Brighton.

Dockery, D. W. (1993) An association between air pollution and mortality in six US cities, *The New England Journal of Medicine*, **329**, 1753–9.

DoE (1992) *The UK Environment*, HMSO, London.

Galloway, J. N. (1989) Atmospheric acidification: projections for the future, *Ambio*, **18**, 161–6.

Hawksworth, D.L. and Rose, F. (1970) Qualitative scale for estimating sulphur dioxide pollution in England and Wales using epiphytic lichens. *Nature*, **227**, 145–8.

QUARG (1993) The Quality of Urban Air Review Group, *Urban Air Quality in the UK*, Department of the Environment, London.

Renberg, I. and Battarbee, R.W. (1990) *Surface Waters Acidification Programme*, pp. 281–300, Cambridge University Press, Cambridge.

Ryaboshapko, A.G. (1985) Sulphur in the biosphere, *Priroda* N7, 42–50.

Schwar, M.J.R. and Ball, D.J. (1983) *Thirty Years On*, The Greater London Council, London.

Sandnes, H. (1993) *Calculated Budgets for Airborne Acidifying Components in Europe*, Meteorological Synthesising Centre, Blindern, Oslo.

UNECE EMEP (1983) Co-operative Programme for Monitoring and Evaluation of Long-Range Transmission of Air Pollutants in Europe, Norwegian Meteorological Institute, Oslo.

Weubles, D.J. (1993) Global climate change due to radiatively active gases, in C.N. Hewitt and W.T.Sturges, *Global Atmospheric Chemical Change*, pp. 53–92, Elsevier, London.

WHO (1987) *Air Quality Guidelines for Europe*, World Health Organization, Copenhagen.

Chapter 8

Georgii, H.W. and Meixner, F.X. (1980) Measurements of tropospheric and stratospheric SO_2 distributions, *Journal of Geophysical Reseach*, **85**, 7433–8.

Stolarski, R.S. in Kaye, J.A. and Jackman, C.H. (1993) Stratospheric ozone change, in C.N. Hewitt and W.T. Sturges, *Global Atmospheric Chemical Change*, pp. 123–68, Elsevier, London.

Turco, R.P., Whitten, R.C., Toon, O.B., Inn, E.C.Y. and Hamill, P. (1981) Stratospheric hydroxyl radical concentrations, *Journal of Geophysical Reseach*, **86**, 1129–39.

Whitten, R.C., Turco, R.P. and Toon, O.B. (1980) The stratospheric sulphate layer, *Pure and Applied Geophysics*, **118**, 86–207.

Chapter 9

Cameron, A.G.W. (1980) Elementary and nuclide abundances in the Solar System. [As cited by Lewis and Prin (1984).]

Gerlach, T.M. and Graeber, E.J. (1985) Volatile budget of the Kilaeau volvano, *Nature*, **313**, 273–7.

Heald, E.F., Naughton, J.J. and Barnes, I.L. (1963) *Journal of Geophysical Reseach*, **68**, 546.

Henderson-Sellers, A. (1983) *The Origin and Evolution of Planetary Atmospheres*, Adam Hilger, Bristol.

Kaufmann, W.J. (1979) *Planets and Moons*, W.H. Freeman, San Francisco.

Mason, B. (1966) *Principles of Geochemistry*, John Wiley and Sons, New York.

INDEX

○ ○